Encounters With Life

Hans F. E. Wachtmeister

Larry J. Scott

Mary A. Perry

Tidewater Community College
Virginia Beach Campus

Morton Publishing Company
295 W. Hampden Avenue, Suite 104
Englewood, Colorado 80110

PREFACE

This laboratory guide was designed for use in a one-semester or a one-year introductory biology course at the college level. Since each exercise is a self-contained unit containing clearly stated objectives, a variety of learning experiences, and thought-provoking review questions, the order of use of the exercises may be varied by the instructor to suit the organization of his course.

TO THE STUDENT: You are encouraged to think for yourself as much as possible. The instructors are more than willing to help you in the lab. If you have any questions, feel free to ask. Sometimes learning is best accomplished by trying, then asking, because until you have attempted an exercise, what you do not understand is not always clear to you.

The exercises are written with the idea that you will read them in advance of the lab period. Most of the exercises are detailed and will require the entire lab period to complete, therefore it is imperative that you be prepared when you begin so that your time may be spent more fruitfully in *doing*. At the end of each exercise is a set of review questions. You are encouraged to complete the answers to these questions before leaving the lab. Some of the questions may require some additional outside reading on your part before you can answer them.

Regular attendance is extremely important for two reasons:

1. If you miss one lab, you are missing the equivalent of one week of work.

2. The materials used in each lab will not be available later, so it is impossible to make up lab work.

Each student is responsible for cleaning up the materials he or she has used during the lab and for returning equipment and materials to their original places. In this way, all students taking the lab will be able to begin with materials and equipment which are clean and in good order, just as you did.

ACKNOWLEDGMENTS: We would like to extend special thanks to our colleagues Sharon Lawhorne, Arlien Steiner, and Agnes Flemming for their contributions in the preparation of this edition of our laboratory guide and in the development of the original exercises upon which it was based. We especially thank John Joyce who put a great deal of effort into this manual.

Special thanks also go to Larry Blevins, Staff Photographer for Learning Resources Center, Virginia Beach Campus, Tidewater Community College for his excellent photographs which grace this laboratory guide and again to Arlien Steiner for the detailed electron micrographs.

Others without whose help this guide could not have been completed include Anna Ferreri, Carol Kasper, Tom Weikel, Jim Hatstat, Donna Wright, Deborah Miller, Andi Helfant, Rita Parisi, Wendell Taylor, and Neil Baynard.

Hans F. E. Wachtmeister
Larry J. Scott
Mary A. Perry

CONTENTS

THE MICROSCOPE AND CELL STRUCTURE

exercise 1

I. Objectives

After the completion of this exercise, the student should be able to do each of the following:

A. Identify, locate, and tell the functions of the main parts of the **compound microscope** and the **stereomicroscope.**

B. Properly carry, care for, and put away both types of microscopes (be able to list steps in putting the microscopes away).

C. Properly examine a slide under the compound microscope.

D. Properly examine a plastomount or small opaque object under the stereoscope.

E. Tell how or if the apparent direction of movement differs from the true direction of movement when observing a specimen under the compound microscope and under the stereoscope.

F. Determine the total magnification afforded by either type of microscope in any given position.

G. Identify on slides and explain the functions of the following cell parts: **cell wall, plastids** (including **chromoplasts, amyloplasts,** and **chloroplasts**), **cell membrane, nucleus,** and **nucleolus**.

H. Determine whether a cell is a plant cell or an animal cell on the basis of its observable structure.

I. Tell how each of the following changes as the magnification changes on a compound microscope: field of view, working distance, and light intensity.

J. Properly prepare and observe a wet mount under the compound microscope.

K. Tell the main advantage(s) of each of the following over the compound light microscope: **phase contrast microscope, transmission electron microscope,** and **scanning electron microscope.**

L. Compare and contrast the use of the transmission electron microscope with that of the scanning electron microscope and be able to distinguish between pictures taken by each.

M. Locate a small animal or alga in pond water using the stereomicroscope. Transfer it to a slide with coverslip. Observe it under the compound microscope, estimate its size, sketch it, and identify it using such resources as the lab manual or other books provided by the instructor.

N. Answer the review questions at the end of this exercise.

II. Compound Light Microscope

A. Care and Use

Much of the work in a biology lab involves the use of microscopes. With them one can open a window into new worlds. However, microscopes are quite expensive ($600 - $800) and should be handled with great care. If something goes wrong with the microscope, notify the laboratory instructor. Do not attempt to fix it or force any mechanical part.

1. Obtain a compound light microscope from the cabinet. Note that the number on the microscope corresponds with the number of the cabinet cubicle. Carry the microscope in an upright position with one hand holding the arm and the other supporting the base. Place it on the lab table in front of you, approximately four inches from the edge. Now get a prepared slide designated by the instructor.

2. Label the illustration of the compound microscope with the following terms:

   ocular — lens nearest the eye
   body tube — keeps ocular and objective lenses at proper distance from each other
   nosepiece — permits interchange of low, medium, and high power objectives
   pointer — found in ocular, can be moved by turning the black eyepiece
   arm — supports body tube and adjustment knobs
   objectives — contain lenses of various magnifications
   coarse adjustment — moves objectives
   fine adjustment — permits exact focusing
   base — bears the weight of microscope
   stage — supports slides
   stage clips — hold slide steady
   iris diaphragm — regulates amount of light going through specimen
   illuminator — provides light source

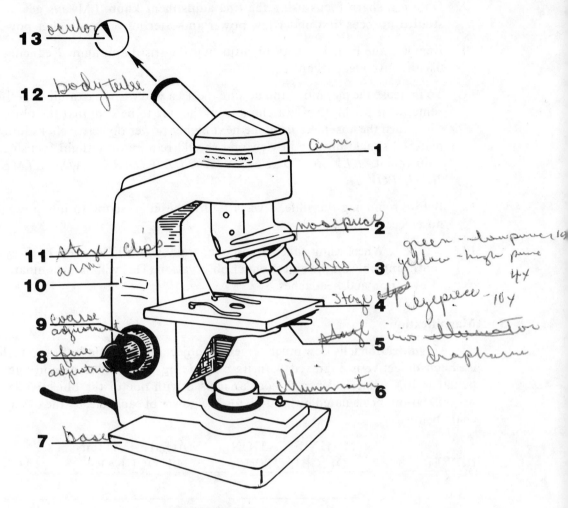

13 — ocular

12 — body tube

11 — stage clip / arm

10 —

9 — coarse adjustment

8 — fine adjustment

7 — base

1 — arm

2 — nosepiece

3 — lens

4 — stage clips

5 — diaphragm

6 — illuminator

green - low power
yellow - high power
4x
eyepiece - 10x
iris illuminator
diaphragm

FIGURE 1.1
COMPOUND LIGHT MICROSCOPE

3. Clean all of the exposed lenses at the beginning of each laboratory with special **lens paper.** *(DO NOT USE PAPER TOWELING, KIMWIPES, OR CLOTH* as this will scratch the lenses.) If the view through the microscope becomes blurred, additional cleaning with lens paper may be necessary.

4. Unwind the lamp cord and insert the plug in the nearest electrical socket.

5. Make sure that the low power objective is clicked into position under the body tube. Turn the substage switch on.

6. Place the slide (coverslip up) on the stage and center the specimen over the opening in the stage.

7. Watching from the side, use the coarse adjustment knob to move the lowest objective as close to the slide as it will go (it should not touch the slide).

8. Next, while looking through the ocular, use the coarse adjustment knob to slowly move the objective upward until the material comes into focus. It should do so in less than a full turn. If it does not, check to see that the material is centered on the stage, lower the objective (watching from the side) and try again. (Always focus moving UP, never down.) Learn to keep both eyes open when looking through the microscope; this prevents eyestrain and headaches.

9. Obtain a sharp focus using the fine adjustment knob. (Always get the object to be studied in focus first under low power and later move to a higher power.)

10. Regulate the light intensity by adjusting the iris diaphragm. Reducing the light will usually give greater contrast.

11. To increase the magnification be sure the area you wish to examine specifically is in the center of the field; then, watching from the side to be sure that the objective clears the slide, turn the nosepiece until the next higher power objective clicks into position. The material should now be in view and should require only slight focusing with the fine adjustment. *NEVER FOCUS WITH THE COARSE ADJUSTMENT UNDER HIGH POWER.*

12. Before removing the slide, always return the microscope to low power and turn the nosepiece all the way down.

13. *NOTE:* When using this microscope, train yourself to keep both eyes open, concentrating only on the object under observation. This will help eliminate the possibility of eyestrain and headaches.

B. Magnification

Magnification tells how much larger the object appears under the scope than it actually is. Each objective is engraved with its magnifying power. The magnifying power of the ocular is 10X (times). When the ocular is used with one of the objective lenses, the **total magnification** is calculated by multiplying the power of each lens by the other. Complete the chart below.

POWER	MAGNIFICATION OF OBJECTIVE	MAGNIFICATION OF OCULAR	TOTAL MAGNIFICATION
Low	_____	_____	_____
Medium	_____	_____	_____
High	_____	_____	_____

C. Field of View

The circular field you see when you look through the ocular is called the **field of view.** The field of view changes in size at different magnifications. At what magnification can you view the largest area on the slide? _____ At what magnification is the field of view the smallest? _____

D. Optional: Measurement with the Compound Microscope

The metric system provides convenient units for expressing small measurements often used with the microscope. The most useful units are millimeters and micrometers. One micrometer (μm) = 1/1000 millimeter (mm) and 1 millimeter = approximately 1/25 inch. 1 micrometer = _____ inch.

In order to estimate sizes of microscopic objects, it is necessary to know the size of your field of view. Using a plastic ruler in the same fashion as you would a slide estimate the diameter of this field on low and medium power. Since the ruler is too thick and the markings too far apart to measure the diameter of the high power field, you will have to use another type of ruler. Measure an object such as a brine shrimp under medium power, and then **use the organism as your ruler.**

At low power the diameter of the field of view is _____ mm.

At medium power the diameter of the field of view is _____ mm.

The brine shrimp is _____ mm. long.

At high power the diameter of the field of view is _____ mm.

E. Depth of Focus

The **depth of focus** is the thickness of an object which is all in sharp focus at the same time. On a microscope this is fairly thin, so, to adequately view a thick specimen, you will have to focus up and down over its body. Pick out a slide with three colored threads, which have been mounted at varying depths. Use the compound microscope to determine the order of threads. *REMEMBER:* No two threads will be in focus at the same time. Not all of the slides are mounted in the same order, so check with the instructor if you are not sure. Medium power will probably help you most, along with a dim light.

Bottom Color _____ Middle Color _____ Top Color _____

F. Illumination and Working Distance

The **working distance** (space between slide and objective) and the light intensity change as you switch objectives. Focus on a slide, and without adjusting the diaphragm, switch from low to high power and compare these qualities.

Does the working distance increase or decrease as the magnification increases?_____

Does the light intensity increase or decrease as the magnification increases?_____

G. Type of Slides

1. Prepared slides have been made in advance by time-consuming techniques. Living material is first killed and cleared by solvents, embedded in paraffin wax, and often sliced into thin sections onto a slide. Then the paraffin is removed from the tissue by solvents and the sliced sections are stained by one or more permanent dyes. Balsam glue is placed over the tissue and a thin glass cover slip is carefully put over it.

   The label often indicates the name of the organism, the particular part or structure of that organism, and the type of slide, if any. (Note these key abbreviations: **w.m. - whole mount, x.s. or c.s. - cross section,** and **l.s. - longitudinal section.**)

   Prepared slides are expensive, so treat them carefully. Never stack one slide on top of another. Always note the label on the side and put the slide back in the designated box.

2. When living tissue or cells are desired for microscopic investigation, **wet mounts** are often made in the lab such as the following:

   Pick a leaf from an *Elodea* plant and place it on a blank slide in a drop of water. Slowly lower the coverslip from one side of the water drop so that any air bubbles present will be pushed out as the coverslip comes down. There should be enough water to reach all edges of the coverslip, but not so much that it leaks out on the slide.

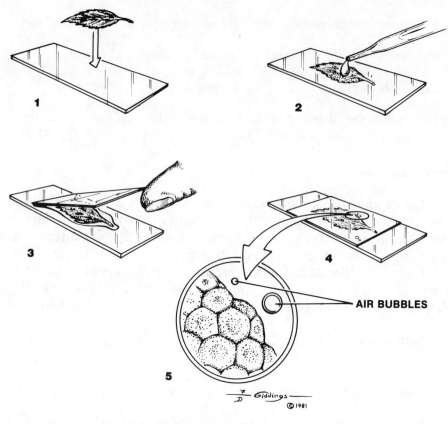

FIGURE 1.2
PREPARING A WET MOUNT

III. Plant Cells

A. Elodea Cells

Look closely at your wet mount of *Elodea* under high power and pick out a single cell. All plant cells are surrounded by a **cell wall** on the outside of the **cell membrane.** The cell walls in unstained cells appear as glassy lines. The cell membrane is held tightly against the cell wall and is nearly impossible to detect under normal conditions. On the inside of the cell membrane is the complex colloid known as **protoplasm.** Sometimes it is subdivided into the **cytoplasm** (all protoplasm except inside the nucleus) and **nucleoplasm** (that inside the nucleus). The cell's inner packages float in the protoplasm.

Such packages are called **organelles**; the most prominent of which is the **chloroplast**, in *Elodea*. These are the green ovals which may be seen flowing around the periphery of the cell as **cyclosis (cytoplasmic flow)** occurs. At some point the chloroplasts will appear to pile up as if held back by a barrier. If you look closely, you may be able to see the greyish ghostlike **nucleus** in the middle of the pile.

The largest, although most transparent organelle is the **vacuole** — a "bag" of water and minerals which occupies most of the center of the cell. This organelle is bound by a membrane known as the **tonoplast.** When plant cells lose water and shrink, it is the vacuole which loses most of the fluid.

6

Draw one *Elodea* cell. Label the cell wall, chloroplasts, and cytoplasm.

B. Onion Leaf Epidermis

An onion bulb is actually a collection of food storage leaves. Peel the filmy epidermis from the bulb leaf and mount in tap water. Dim the light and find the **nucleus** — much easier to see here, since there are no chloroplasts. Close examination should reveal one or more **nucleoli** in the nucleus. Also note the movement of small particles in the nearly transparent cytoplasm. This is Brownian movement, which will be discussed in a later exercise. Stain may be used if you have difficulty locating the nucleus or nucleoli. Draw a few onion cells with a nucleus in the space below.

C. Carrot Root

Shave a thin slice of carrot onto a slide and prepare a fresh mount of it. Observe the orange-yellow bodies called **chromoplasts,** another type of plastid. Draw a few of the cells with chromoplasts in the space below.

D. Cells of the Potato

White potatoes are stems inflated with starch, stored within the cells in membrane-bound packages (**amyloplasts**) containing starch grains. Cut a wedge-shaped sliver of potato. (This ensures that one edge will be thin enough to see through.) Add a drop of iodine, a coverslip and observe under the microscope.

By dimming the light with the diaphragm lever, you may be able to see the faint growth rings in some of the grains and the nearly transparent cell walls, which look rather like soap bubbles. The staining may take a few minutes, since the iodine must diffuse through both the cell membrane and the amyloplast membrane.

Draw a few of the potato cells with amyloplasts. Label cell wall and starch grains.

IV.   Animal Cells

Although animal cells do not have cell walls or plastids, they have many other parts in common with plant cells. (Refer to your textbook for a comparison between plant and animal cells.)

A.   Human Cheek Cells

The cells on your skin are constantly sloughing away. In most areas they dry as they detach; such cells are not helpful in studying structure, since this breaks down as the cells dry. Such surface-coverings, composed of **epithelial cells,** are also formed on the inner linings of your cheeks. The cells here are still moist and will show structure if collected.

Rub the inner side of your cheek with a toothpick to pick up some cells. Scrape the cloudy (cell-containing) fluid onto a slide, add a drop of dilute methylene blue, and observe.

These cells function only as shielding, and therefore have no elaborate internal structure, but you should be able to identify the **cell membrane** (no cell wall, you notice) of the pancake-shaped cells, the thin **cytoplasm, nucleus** and **nucleolus.** Nucleoplasm will stain darkest. Draw one or two epithelial cells in the space provided on the next page.

B. **Human Blood: Stained**

Wright's stain is used to make white blood cells more visible among the **red blood cells.** Obtain a prepared slide of human blood and see if you can recognize the RBC's. There are several types of **white blood cells** (WBC's), all larger than the RBC's, with differing shaped purple-stained nuclei. See how many different types of WBC's you can find.

You will have to look for these, since there are only 1 or 2 for every 1000 RBC's in normal blood. A person whose body was fighting infection or who had a blood disease such as leukemia would produce more WBC's.

Among all of the cells are scraps of cells originating in the bone marrow — the **platelets.** These greyish scraps are very important in the initiation of blood clotting. They have no nuclei; being only cell fragments. (The red blood cells have no nuclei either, but theirs have disintegrated.) Draw a few red blood cells, a white blood cell, and a few platelets in the space provided below.

V. Stereomicroscope

The **stereomicroscope** (also known as the **dissecting scope** or **binocular scope**) has two oculars. Although magnification is much lower than with a compound light microscope, the depth of focus is similar to that of our eyes. Also, transmission of light through the object is not always necessary since there are two sources of illumination; a substage light and an above-stage light. Thus, thick, opaque objects can be viewed with this microscope.

A. Obtain a stereoscope from the cabinet using the same care as you did with the compound microscope.

B. Label the following illustration. Use the terms in the column on the right.

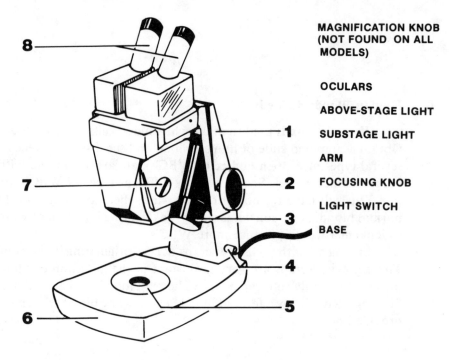

MAGNIFICATION KNOB (NOT FOUND ON ALL MODELS)

OCULARS

ABOVE-STAGE LIGHT

SUBSTAGE LIGHT

ARM

FOCUSING KNOB

LIGHT SWITCH

BASE

FIGURE 1.3
STEREOMICROSCOPE (DISSECTING MICROSCOPE)

C. Get a fingerbowl with a small preserved animal in it. Following the directions below, examine it under the stereomicroscope.

D. Place object on stage and center the area to be examined.

E. If there is a magnification knob, turn it to the low power. The total magnification for low is 10X; high power is 20X. If there is no such knob, the microscope magnifies 15X. For material that is **transparent,** the substage light will work best; if the object is **opaque,** use the overhead or above-stage light. (It is seldom advisable to use both lights at once.)

F. Focus by turning the large, black focusing knob.

G. If your scope has two powers, you may want to switch it to high power to look at the object more closely.

H. Try different kinds of objects on the stereomicroscope to see what this piece of equipment is capable of, e.g., butterfly wing, housefly, fingertip, plastomount, dish of pondwater, etc.

VI. Other Types of Microscopes

A. In each lab there is one **teaching** or **"team"** scope. This is just like the compound scope you use, except that there are two oculars, so two people can look at the same thing at the same time. There is also a green arrow pointer which can be moved by means of a knob on the body tube.

B. There is also one **phase-contrast microscope** in each lab. This scope uses regular light rays for illumination, but, by filtering out certain rays, allows the observer to see objects (particularly tiny moving objects such as cilia and flagella) more distinctly. This microscope is especially good for observing live material.

C. In most cases, material observed under electron microscopes is first killed and fixed in a condition as close to the living one as possible. Extremely thin sections of the material are made for viewing under the **transmission electron microscope,** which is able to increase magnification and resolution (clarity) by passing electrons, rather than light rays, through the specimen. Huge electromagnets are used to spread the electrons, in place of glass lenses used to spread light rays. Small organisms or groups of cells may be observed in the **scanning electron microscope,** which uses a moving beam of electrons to bounce electrons off of the specimen. These are detected electronically and an image is produced on a viewing screen. The biggest advantage of the scanning electron microscope is its great depth of focus, coupled with high magnification.

D. Observe the following scanning and transmission electron micrographs and those available in the lab. Also look at the material on demonstration under the phase-contrast and teaching microscopes.

PIECE OF HUMAN HAIR
(750X)

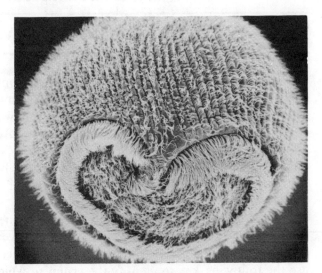

STENTOR
(500X)

FIGURE 1.4
SCANNING ELECTRON MICROGRAPH

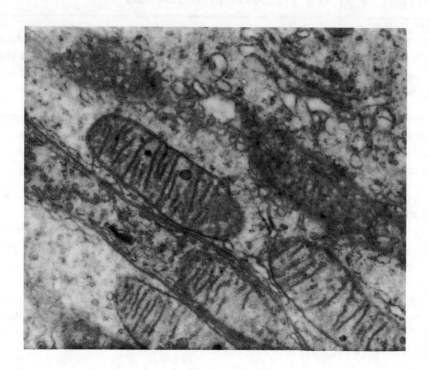

MITOCHONDRIA IN A LIVER CELL
(18,7000X)

FIGURE 1.5
TRANSMISSION ELECTRON MICROGRAPH

VII. A Window Into A New World

If you feel confident of handling a microscope properly, obtain a small dish with pond or aquarium water. Use the stereoscope to locate any small animals, plants, rocks, etc. Then use a small medicine dropper to collect some of this water and make a wet mount with a slide and coverslip. Use the compound light microscope to view the small organisms. Dim the light. The light creates heat which can kill small animals in a short time. You will find that they move rather quickly out of the field of view. It will take a bit of practice to follow an organism with a microscope which turns all images upside down and backwards — if you want to go left, push the slide right, etc.

The following illustrations include some of the most commonly observed organisms usually found in our aquariums and local ponds. See how many you find, or, if you see one not pictured, draw it, to give you practice in drawing from a microscope. Your instructor will show you how to use the following key to identify the group to which the organisms you see belong.

Note: Organisms are not drawn to scale.

After you have located an interesting organism, attempt to "key it out", that is, identify the specimen by use of a key.

A **key** is an arrangement of characteristics of species which aids in their identification. While a key can never include **all** the organisms of an area, it should enable one to identify most of the commonly-encountered species.

This key, by which you will identify major groups of pond organisms, is referred to as **dichotomous.** That is, one needs only to make a single choice — whether the unknown has this particular trait or doesn't.

So, while observing one particular organism, follow the first couplet and make the choice. Then go to the next couplet as directed by the key. Once you believe you have identified the organism, tell your instructor its group name. Your instructor may ask you to tell how you worked through the key.

KEY TO ORGANISMS COMMONLY FOUND IN FRESH WATER

1a.	Cells submicroscopic (usually appearing as dots, dashes, or commas under high power. May be clustered or in chains)	Bacteria
1b.	Cells not submicroscopic	2
2a.	Organisms green	3
2b.	Organisms not green	10
3a.	Pigment blue-green, dispersed evenly in cells; gelatinous sheath often present	Blue-green Algae
3b.	Pigment not blue-green, pigment usually found in plastids	4
4a.	Pigments brown, golden brown, or greenish yellow	5
4b.	Pigments leaf-green	6
5a.	Cells non-flagellate, cell walls decorated with striations, bumps, and grooves	Diatoms
5b.	Cells non-flagellate, pigments greenish-yellow, cell walls not decorated	Yellow-green Algae
5c.	Cells flagellate, pigments golden brown, cell walls not decorated	Golden-brown Algae
6a.	Cells with flagella or cilia	7
6b.	Cells without flagella or cilia	8
7a.	Cells inflexible — always keep same shape	Flagellated Green Algae or Spores of Green Algae
7b.	Cells flexible — can change shape a little as they move	Euglenoids
8a.	Cells arranged in colonies (chains, balls, sheets, etc.)	Green Algae
8b.	Cells usually not arranged in colonies	9
9a.	Cells divided into 2 equal parts by constriction	Desmids
9b.	Cells not divided into 2 equal parts, cell shape usually spherical or ovoid	Unicellular Green Algae
10a.	Organism unicellular or colonial (protists)	11
10b.	Organism multicellular (organs of digestion usually visible inside animals)	12

11a.	Cilia or flagella present, pseudopodia absent	13
11b.	No cilia or flagella, pseudopodia fingerlike or pointed in radially symmetrical forms. May have a shell	Sarcodines
12a.	Body shape wormlike	14
12b.	Body shape not wormlike	15
13a.	Cells with cilia	Ciliates
13b.	Cells with long whiplike flagella	Flagellates
14a.	Body segmented	Bristle Worms
14b.	Body unsegmented, movement by S-shaped lashing	Nematodes
15a.	Jointed appendages present	16
15b.	Jointed appendages absent, forked foot, cilia around mouth	Rotifers
16a.	Have 2 pr. antennae and/or legs on abdomen	Crustaceans
16b.	Have 1 pr. antennae, no legs on abdomen	Insects

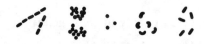

BACTERIA

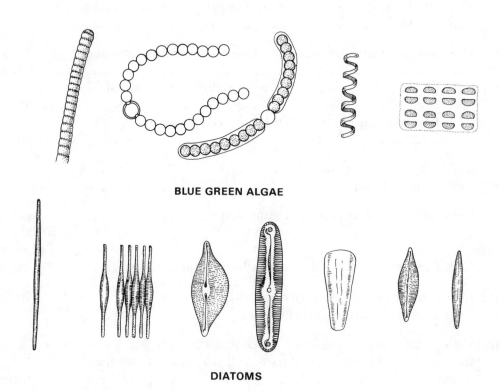

BLUE GREEN ALGAE

DIATOMS

FIGURE 1.6
MICROSCOPIC ORGANISMS COMMONLY FOUND IN POND WATER

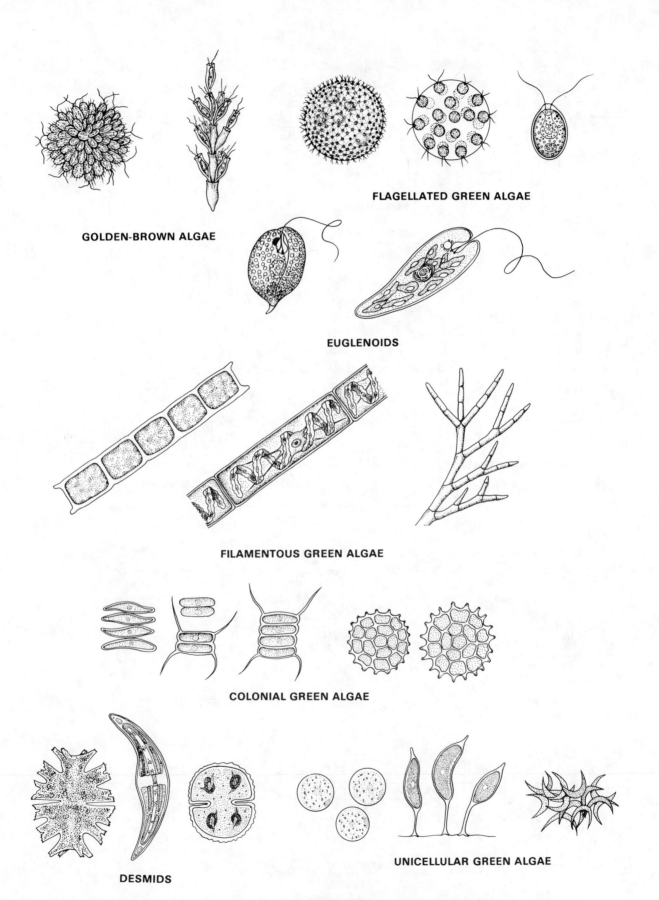

GOLDEN-BROWN ALGAE

FLAGELLATED GREEN ALGAE

EUGLENOIDS

FILAMENTOUS GREEN ALGAE

COLONIAL GREEN ALGAE

DESMIDS

UNICELLULAR GREEN ALGAE

FIGURE 1.6
MICROSCOPIC ORGANISMS COMMONLY FOUND IN POND WATER

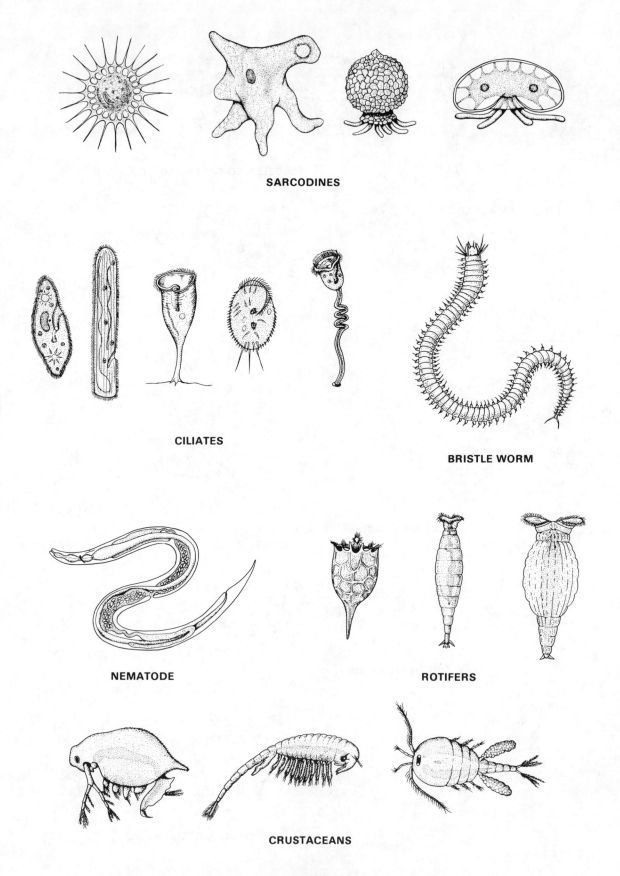

SARCODINES

CILIATES

BRISTLE WORM

NEMATODE

ROTIFERS

CRUSTACEANS

FIGURE 1.6
MICROSCOPIC ORGANISMS COMMONLY FOUND IN POND WATER

VIII. Putting Away and Storage of Microscopes (All Types)

A.   Remove the slide or object being observed and be sure all parts are dry.

B.   Set the microscope on the lowest magnification possible and lower the objectives.

C.   Turn off the light and wind the cord **loosely** around your hand, then loop the plug through the coil and place the coil on the microscope stage. Cover the microscope with its plastic dust cover.

D.   Put the microscope back in the proper cubicle in the cabinet. Compound microscopes have **numbered** cubicles; stereomicroscopes have **alphabetized** cubicles.

IX. Review Questions

A.   What is the function of the pointer on a microscope? _____
_____

B.   The total magnification of 100X is achieved under what objective: low, medium, or high?
_____

C.   What is meant by the working distance of a microscope? _____
_____

D.   Which of the four types of microscopes found in our lab does not invert the image? ____

E.   Which microscopes do not use beams of light? _____

F.   What do the following abbreviations mean?   w.m. _____

     c.s. _____ and l.s. _____

G.   What does the term "opaque" mean? _____

H.   What is a wet mount? _____

I.   What cell parts were seen in the plant cells? _____
_____

J.   What cell parts were seen in the animal cells? _____

K.   Name a human cell that does not have a nucleus. _____

EXCHANGE BETWEEN CELLS AND THEIR ENVIRONMENT

I. Objectives

After the completion of this exercise, the student should be able to do each of the following:

A. Tell what causes **Brownian movement**. Be able to recognize it and describe it.

B. Define **diffusion** and tell how it was demonstrated in the lab (include *every* demonstration of diffusion).

C. Define **osmosis** and explain how it was demonstrated in the lab.

D. Use the following terms in describing solutions and their relationships to each other: **solute, solvent, hypertonic, hypotonic,** and **isotonic.**

E. Explain the difference between **hemolysis** and **crenation** of erythrocytes and tell how each may occur.

F. Distinguish between **turgor** and **plasmolysis** of plant cells and explain how each was demonstrated in the lab.

G. Answer the review questions at the end of this exercise.

II. Introduction

For a cell to live and function, certain substances must enter the cell (e.g., food, water, and oxygen) while others must leave the cell (e.g., waste materials). Cellular organization and makeup is so complex that it cannot be easily studied, and is far from being completely understood. However, there are many properties of living cells which can be explained in terms of physical and chemical laws.

Most physical phenomena occurring in cells are consequences of molecules being in constant oscillating motion, which increases as temperature increases. This molecular motion produces a variety of effects: Brownian movement, diffusion, osmosis, facilitated diffusion through cell

19

membranes, and others. Most of these phenomena are extremely important to the functioning of living systems; life is as much a result of specific physical phenomena as a result of specific chemical reactions. Many of these phenomena are easily observed in living and non-living systems.

III. Brownian Movement

Brownian movement is visible under high power of the microscope. Minute particles of solid matter, bacteria and other tiny organisms in a liquid suspension seem to move constantly in an erratic, random course. This motion is due to the bombardment of the larger particles by the molecules in the fluid. Energy which keeps particles larger than a molecule in motion is the kinetic energy of the molecules in the surrounding medium; these strike the larger particle on all sides and impart to it a movement similar to their own kinetic motion. The movement of larger particles is obviously slower than that of smaller ones.

Tap a tiny bit of powdered carmine dye from the tip of a toothpick onto a drop of water on a slide, add a coverslip, and observe the quivering motion of the particles under high power. (A mass movement of particles in any one direction is due to the microscope not being quite level, using only one stage clip, or catching the coverslip under the stage clips. This is not what you are looking for.)

IV. Diffusion

Particles in solution tend to move from areas where they are highly concentrated to areas where they are less concentrated until a uniform distribution of particles is achieved. The phenomenon of particles scattering in this manner is called **diffusion.**

A. In front of the labroom the instructor will open a bottle of a very volatile fluid such as ether, acetone, or turpentine. The fluid molecules will evaporate and escape from the bottle into the air. Note how long it takes your body's sense of smell to detect these gaseous molecules.

What causes the liquid to change into a gas? _____

Which students detected the gaseous molecules first? _____ Last? _____

What might cause irregularities in the distribution of these molecules in the room?

_____

B. When a solid dissolves in water, the molecules or ions scatter throughout the water until their distribution is uniform. This is easy to observe if the dissolving substance is colored.

Fill a small test tube 2/3 full of tap water, then drop a crystal of purple $KMnO_4$ (potassium permanganate) into it. Set the tube in a place where it will not be jostled and observe the dispersal of the colored material throughout the remainder of the lab period.

How might the dispersal be speeded up? _____

_____

C. On a petri dish of agar gel place a crystal of potassium permanganate approximately 5 centimeters from an equal size crystal of malachite green. Note the rate of diffusion. Find the molecular weights of each compound from the jar label.

m. w. of malachite green _____

m. w. of potassium permanganate _____

How does the molecular weight of the compound affect the rate of diffusion? _____ .

V. Diffusion Through A Semipermeable Membrane

A **semipermeable,** or **selectively permeable, membrane** is one that allows some particles (molecules and ions) to pass through it and restricts the movement of others. **Cell membranes** are semipermeable and certain substances can freely diffuse through them without any expenditure of energy by the cell (this is called **passive transport**). The selection of particles by the cell membrane is based on size and other factors.

Passage of water molecules is virtually impossible for a cell to stop. The diffusion of water molecules through a semipermeable membrane is called **osmosis.** If a sugar solution is separated from pure water by a semipermeable membrane, there will be a greater movement of water molecules from the pure water into the sugar solution than vice versa. The water molecules will move from an area of high concentration of water to an area of lower concentration of water. Only the water molecules will diffuse because sugar molecules are too large to pass through most semipermeable membranes (including cell membranes).

Cellophane dialysis tubing is a synthetic semipermeable membrane which permits diffusion of water and other small molecules while restricting passage of large molecules. We will create artificial "cells" with strips of this tubing. (The tubing is flattened when dry. Soak the strip in water for about a minute, then open by rolling between your thumb and fingers.)

A.   For this experiment tie one end of a 20 cm strip of tubing very tightly with string and fill to about 5 cm of the top with syrup or an 80% glucose solution. Insert a 1 ml pipette, and tie the open end shut very securely around the pipette. Support the bag in a beaker of water as shown in the diagram and note the position of the column of fluid for about 25 minutes. (If there is no change after 5 minutes, check for leaks, which appear as wiggly lines in the water. It will take a few minutes to build up pressure in the bag.

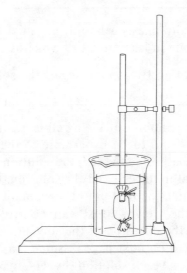

FIGURE 2.1
APPARATUS FOR EXPERIMENT A

B.   Fill a 10 cm strip of tubing with a starch solution and tie off both ends tightly. Rinse with tap water, then submerge in water containing enough iodine (I_2-KI) to approximate the color of beer. When starch molecules touch iodine, a blue or purplish color appears. Observe the bag after about 15 minutes for any color change. Based on the location of the stain, you should be able to answer the following questions.

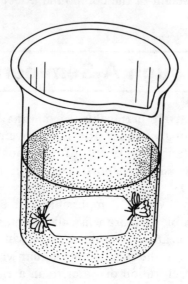

FIGURE 2.2
STARCH-IODINE EXPERIMENT

Could iodine get through the membrane? _____ How do you know? _____

_____

Could starch get through the membrane? _____ How do you know?_____

_____

Which experiment (A or B) demonstrates osmosis?_____

Why is the other experiment **not** a demonstration of osmosis?_____

_____

C.   When two substances (usually one is a liquid) are mixed, the one in greatest quantity is called the **solvent** and the one in least quantity which is dissolved is the **solute**.

With a cup of coffee containing 1 teaspoon sugar, the solvent is _____

and the solutes are _____ and _____ .

When the solute concentration outside a cell is greater than the solute concentration inside a cell, the solution is said to be **hypertonic** to the cell. If the solute concentration outside the cell is less than that inside the cell, the solution is **hypotonic** to the cell, and if the solute concentration is equal in the cell and the solution, the cell is in an **isotonic** solution. Most cells are designed to function best in isotonic conditions, but some have the ability to tolerate solutions unlike their own. Most can control the movement of large solute molecules, but cannot stop the diffusion of the solvent.

In the situations pictured on next page, assume that the solute molecule is too big to penetrate the membrane. Label each solution (hyper-, hypo-, or isotonic) and use an arrow to show the direction of the heavier flow of solvent (water). If the flow in and out is equal, use opposite-facing arrows. The circles indicate cells.

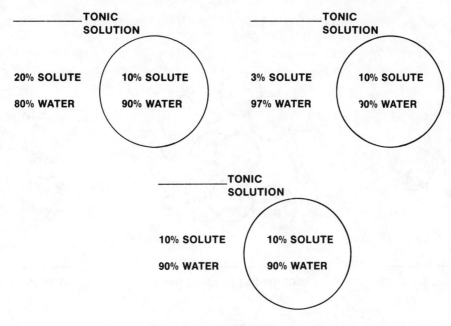

FIGURE 2.3
EFFECTS OF OSMOSIS ON CELLS

D. **Lysis and Crenation**

Red blood cells (erythrocytes) can easily be used to illustrate the osmotic reactions of a living animal cell. The plasma membrane of RBC's is permeable to water but relatively impermeable to salts. If placed in an isotonic saline solution (normal plasma is 0.85% NaCl) it retains its normal shape; if placed in a hypotonic solution, it swells and bursts (**lysis**, specifically known as **hemolysis** in red blood cells) releasing the hemoglobin and producing ghosts (empty cell membranes); when placed in a hypertonic solution, the cells lose water and shrink (**crenation**) showing wavy, irregular outlines.

1. Obtain three clean slides and, with a wax pencil, mark one 5%, one 0.85% and one 0% (distilled water). Set up three microscopes.

2. Place a drop of the appropriate saline solution (NaCl) on each slide.

3. Disinfect your finger with rubbing alcohol and let finger air dry. Prick the finger with a sterile lancet, then throw the cotton and lancet away. Do *NOT* reuse a lancet: dispose of it immediately after use.

4. Touch each slide and deposit a small drop of blood **beside** the drop of saline.

5. Add a coverslip to each slide immediately, raising and lowering the coverslip a couple of times to mix the blood and the saline. Add more of the proper solution at the edge of the coverslip, if necessary.

6. Choose an area where the cells are thinly distributed and observe under high power. The differences are best compared if three microscopes are used at once, side by side.

7. Refer to the previous diagrams to label the following red blood cell diagrams with the proper solution designated and the condition of the cell (hemolyzed, normal, or crenated).

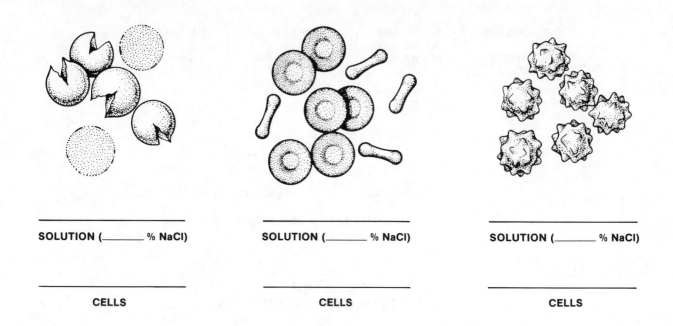

SOLUTION (_____ % NaCl)   SOLUTION (_____ % NaCl)   SOLUTION (_____ % NaCl)

_____   _____   _____
CELLS   CELLS   CELLS

FIGURE 2.4
EFFECTS OF OSMOSIS ON HUMAN RED BLOOD CELLS

E. **Turgor and Plasmolysis**

The **protoplast** (plasma membrane plus all cytoplasm) of a plant cell is confined by a cellulose **wall,** which the plasma membrane secretes. This wall is fairly flexible, but normally plant cells contain enough water to push the cell membrane firmly against the wall, a condition called **turgor.** The rigidity of non-woody plants is due to the turgid condition of individual cells, collectively.

If a plant cell is placed in a hypotonic solution, water will flow into the cell until the wall pressure (resistance of the wall) equals the turgor pressure (osmosis causing the protoplast to swell outward). For this reason plant cells do not rupture as animal cells do in hypotonic situations. It is virtually impossible to tell whether a cell is in an isotonic or hypotonic solution by just observing the cell, but in hypertonic solutions, plant cell protoplasts lose water and shrink **(plasmolyze).** This is analagous to crenation in animal cells.

If you want to make lettuce salad crisp, would you put it in salt, or tap water? _____

_____

Mount a leaf of *Elodea* in water on a slide and observe. These leaves are two cells thick, so you should be able to focus up and down to see that the cells in one layer are larger than those in the other. When one layer is in focus you may be able to see the shadowy outlines of cell walls in the other layer.

As the cells warm, the cytoplasm will begin to flow faster, carrying the green egg-shaped chloroplasts and the nearly transparent nucleus around the cell. Do the chloroplasts and cytoplasm flow uniformly all over the cell or around the edges? (Don't let the other layer's

wall shadows fool you.) _____

Blot off the tap water and add salt solution (5% NaCl) by adding the salt at one side of the coverslip and drawing it under by using a paper towel to absorb the fluid from the opposite side of the coverslip. Let sit for a minute or two and observe again. The cell contains a **vacuole** which holds most of the cell's water, similar to a water bag inside the cell. Note the clear space forming between the cell wall and the cell membrane.

Has water left or entered the cell in the salt solution? _____

Rinse the salt solution from the leaf and remount in tap water. Observe periodically for about 5 minutes.

How is the cell changing?_____

Account for this change in terms of osmosis and the characteristics of the solution.

_____

VI. Review Questions

A. Which is the largest molecule, starch, iodine, or water? _____

How do you know?_____

_____

B. Does Brownian movement occur in the cytoplasm of living cells? _____

_____

C. Would the salt in the ocean be considered the solute, or the solvent? _____

_____

D. A red blood cell is dropped into a 2% salt solution. In relationship to the cell's contents, this solution is:
        a. hypertonic
        b. hypotonic
        c. isotonic

E. Do plant cells have cell membranes? _____

F. Turgor occurs in plant cells, but not in animal cells, due to the presence of a _____ which offsets osmotic pressures.

G. Crenation occurs in cells in a solution which is:
        a. hypertonic
        b. hypotonic
        c. isotonic

CHEMICAL ASPECTS OF LIFE

I. Objectives

After the completion of this exercise, the student should be able to do each of the following:

A. Define **organic** and **inorganic** compounds and describe a simple test to distinguish one from the other.

B. Name the elements most often found in living organisms.

C. Describe the 3 major groups of organic compounds which compose living organisms and give examples.

D. Distinguish between the 3 types of **carbohydrates,** and give an example of each.

E. Define: **polymer, monomer, dehydration synthesis,** and **hydrolysis.**

F. Define **reducing sugar, emulsion, emulsifier, fats, oils, amino acid, glycerol,** and **fatty acid.**

G. Explain how cheese is produced. Use the following terms: **curd, rennilase, enzyme, whey, catalyst,** and **coagulation.**

H. Explain the results of each experiment using each of the following: Benedict's reagent, iodine, and Millon's reagent.

I. Answer the review questions at the end of this exercise.

II. Introduction

All living organisms are composed of compounds which are usually divided into two groups, organic and inorganic. It was once believed that organic compounds were produced only by living organisms. However, many of the thousands of organic compounds are now being made in laboratories. Therefore, a standard definition of an **organic compound** is one whose molecules contain carbon in the form of chains or rings; thus, a compound which does not contain carbon arranged in chains or rings is considered **inorganic.**

Most cells are about 85% water; the bulk of their dry weight consists of carbon (C), hydrogen (H), oxygen (O), nitrogen (N), and phosphorus (P), variously arranged into four major types of organic compounds: **carbohydrates, lipids, proteins,** and **nucleic acids.** The more complex members of these categories are made up of chains of smaller molecules **(monomers)** strung together more or less like beads in a necklace. Such complex molecules are called **polymers.** In living organisms, polymers are made by **dehydration synthesis,** the loss of a water molecule between each pair of monomers. Conversely, polymers can be digested (broken up into monomers) by the addition of a molecule of water between each pair of monomers. This process is known as **hydrolysis.**

III. Test for Organic and Inorganic Compounds

Heat the test tubes, each containing a small amount of one of the substances below, over an open flame until all reactions inside the tubes seem to stop. Note the residue in each test tube. If the deposit is black, the substance left on the test tube wall is carbon and the compound was probably organic. Fill in the chart below.

	Organic	Inorganic
sugar	_____	_____
meat	_____	_____
table salt	_____	_____
baking soda	_____	_____
unknown	_____	_____

IV. Carbohydrates

The approximate ratio of elements in carbohydrates is 1 C: 2 H: 1 O. Simple sugars, or **monosaccharides,** are the simplest of the carbohydrates. There are many types of monosaccharides, but the most prevalent in living organisms is **glucose.**

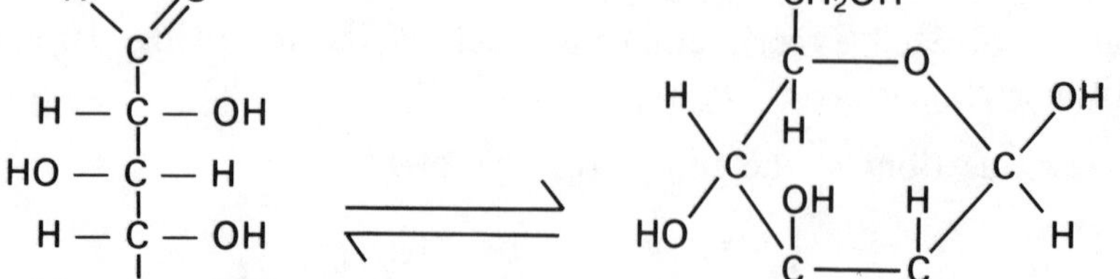

FIGURE 3.1
TWO FORMS OF GLUCOSE IN SOLUTION

A. Reducing Sugars

Sugars which consist of two monosaccharides linked together are called **disaccharides.** When monosaccharides join, it is in the closed position, pictured above. All monosaccharides and some disaccharides have the ability to add electrons to (reduce) other

28

molecules. These sugars are called **reducing sugars.** Reducing sugars have free **carbonyl** groups ($- C \overset{\nearrow O}{-}$ or $- C \overset{\nwarrow O}{\underset{-H}{}}$) in close proximity to **hydroxyl** groups (– OH), and this is where the electrons come from. Frequently, when two monosaccharides join to form a disaccharide, the carbonyl group gets tied up in the linkage, so it is not available to release electrons. This is the case with some disaccharides and all **polysaccharides** (long polymers of monosaccharides).

Benedict's reagent, which is used in testing for reducing sugars, contains copper ions in alkaline solution. The blue color of the reagent is characteristic of solutions containing copper ions. When Benedict's reagent is heated in the presence of a reducing sugar, the copper ions are reduced to metallic copper. This forms a **precipitate** (a substance which settles out of solution) which colors the contents of the tube green to brick red or brown, depending on how much reducing sugar is present. Before the development of paper test strips, this test was used by diabetics to test for sugar (in this case, glucose) in the urine.

1. Since all solutions in this test must be heated, prepare a hot water bath by setting a beaker about ½ full of tap water on a hotplate. Once steam is rising, you can turn the temperature back to medium.

2. Number test tubes 1-9 with a wax pencil and place in a test tube rack.

3. Add the materials to the first 8 tubes indicated in the chart. Mix well.

4. Place in the hot water bath for 2 minutes, remove, cool, and record the color (in your own words) on the chart.

5. In tube 9 combine the sucrose and acid and heat in the water bath for 10 minutes. **Then** remove from heat and add the Benedict's to tube 9. Reheat, cool, and record the results.

TUBE	TEST MATERIAL	BENEDICT'S	OBSERVATIONS	TEST RESULTS (+ or -)
1	tap water (control) 10 drops	5 ml		
2	glucose 10 drops	5 ml		
3	milk 10 drops	5 ml		
4	applesauce 10 drops	5 ml		
5	starch 10 drops	5 ml		
6	molasses 10 drops	5 ml		
7	sucrose 10 drops	5 ml		
8	10% HCl (control) 8 drops	5 ml		
9	sucrose 10 drops 10% HCl 8 drops	5 ml		

B. **Starch**

Starch is a polysaccharide consisting of many glucose molecules linked together. Iodine (I_2-KI) is added to detect the presence of starch. The iodine molecules get stuck in the spirals of the starch molecule and cause a blue-black color to appear. A red color is an indication of the presence of certain **dextrins,** intermediate products of starch digestion whose spirals are shorter.

1. Thoroughly clean the tubes from the previous experiment.

2. Add the materials to the first seven tubes as indicated in the chart. No heating is needed to produce a reaction.

3. Observe and record the results.

4. To the eighth tube, add 1 ml of saliva or bacterial amylase to 1 ml of starch solution. Let this sit for 10-15 minutes, then pour half of the contents into a clean test tube.

5. Add iodine to one of these tubes and test the other with Benedict's solution.

TUBE	TEST MATERIAL	IODINE	OBSERVATIONS	TEST RESULT (+ or -)
1	2 ml starch	5 drops		
2	2 ml glucose	5 drops		
3	2 ml water	5 drops		
4	2 ml sucrose	5 drops		
5	cellulose (cotton)	5 drops		
6	potato (small piece)	5 drops		
7	bread (small piece)	5 drops		
8	starch-saliva	5 drops		

From the results of testing tube 8, what did the saliva do to the starch spirals?_____

_____

What did the amylase convert the starch to?_____

V.   Lipids

Lipids are oily or waxy compounds. They are generally insoluble in water, but are soluble in organic solvents, such as ether, acetone, carbon tetrachloride, and chloroform. The largest class of lipids, the **triglycerides,** are composed of two kinds of component molecules, **fatty acids,** and **glycerol** (an alcohol). Triglycerides which are liquid at room temperature are called **oils,** while those which are solid at room temperature are called **fats.**

```
        H
        |
 H  —  C  —  OH
        |
 H  —  C  —  OH                     OH—C—(CH₂)ₙ—CH₃
        |                               ‖
 H  —  C  —  OH                         O
        |
        H
```

GLYCEROL   **FATTY ACID**

FIGURE 3.2
COMPONENTS OF A TRIGLYCERIDE

Lipids are very important to living systems because they serve as a concentrated source of stored energy (nearly twice as rich in calories per gram as carbohydrates or proteins), and, in entire organisms as thermal insulation and shock absorbing pads for organs, bones, and muscles.

A. **Sudan Stain Test For Lipids**

1. Add 3 drops of vegetable oil to a test tube half filled with tap water.

2. Shake thoroughly and observe the way the oil is only temporarily dispersed. This is an **emulsion,** a mixture of two liquids, each insoluble in the other.

3. Now add a small amount of lipid-specific red Sudan stain and mix again.

4. Add several droppers full of a liquid detergent to the tube and shake again. Allow to stand, noting that the two phases (oil and water) are no longer distinctly separated. Detergent is often termed an **emulsifier.** Its molecules are water-soluble on one end and lipid-soluble on the other. These surround small oil droplets, water-soluble end out, and allow the droplets to stay suspended in the water.

   Which layer stained red with the Sudan? _____

   How did the detergent affect the continuity of the lipid layer? _____

   _____

B. **Greasy Spot Test For Lipids**

   Place a drop of oil and a drop of sucrose solution each on a piece of paper. After 15 minutes, hold the paper toward the light. A transparent to translucent spot is a positive test for a lipid.

VI. Proteins

Proteins are polymers made up of **amino acids** linked by peptide bonds. The "R" shown on the amino acid structural formula shown below can be one of 20 different attachments. Therefore, there are 20 amino acids commonly found in living systems.

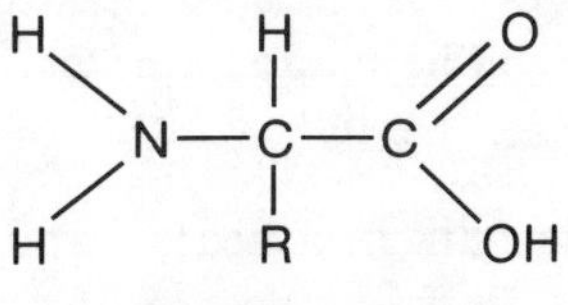

```
   H           H        O
    \          |       ⫽
     N  —  C  —  C
    /          |       \
   H           R        OH
```

FIGURE 3.3
AMINO ACID

The number of ways in which 20 different subunits can be arranged in a chain is extremely large, consequently there is a greater variety of protein molecules in living systems than of any other kind of molecule. And the functions which proteins perform are just as varied.

A. Test for Protein

The "R" on some amino acids is a ring structure. Proteins which contain this type of amino acid (and most do) will turn yellow or pinkish-red when heated with Millon's reagent, which is originally a clear colorless liquid. This is due to a reaction between the nitric acid in the reagent and the ring structures on the amino acids. (CAUTION: MILLON'S RE-AGENT IS VERY CORROSIVE BECAUSE OF THE STRONG NITRIC ACID IT CONTAINS)

1. Clean out test tubes used previously and number 1-7.

2. Add the materials to these tubes indicated in the chart.

3. Place in the hot water bath for 5 minutes, remove, and record observations.

TUBE	TEST MATERIAL	MILLON'S	OBSERVATIONS	TEST RESULT (+ or –)
1	tap water 5 ml	10 drops		
2	sucrose 5 ml	10 drops		
3	albumin 5 ml	10 drops		
4	milk 5 ml	10 drops		
5	bread 1 chunk	10 drops		
6	ground peanuts small amount	10 drops		
7	vegetable oil 5 ml	10 drops		

B. Enzymes

Enzymes are proteins which act as **catalysts** (substances which regulate the rate of chemical reactions, but which do not permanently unite with the reactants).

Milk (as shown by your results with the Millon's reagent) contains a considerable amount of protein. The human stomach produces **rennin,** which converts this milk protein (**casein**) to insoluble **paracasein,** a shorter chainlength product of casein hydrolysis. This precipitates out of solution forming a **curd** and leaves the water and other dissolved substances in the milk behind as a watery **whey.** About three-fourths of the world population cannot produce rennin, and consequently, cannot digest milk protein as easily as those who have it.

In this experiment, cheese will be made from whole milk by the use of a similar enzyme (**rennilase**) produced by bacteria.

1. Put ½ cup or 125 ml whole milk in a beaker and warm to 32° C (88-90° F).

2. Add 3 drops of rennilase to the warmed milk and stir. Remove from heat and allow to stand undisturbed for 15 minutes.

3. By now a curd has been formed. Break up the curd and filter to remove the whey, using a piece of cheesecloth.

4. Salt to taste and press the curd between cheesecloth or paper towels.

5. For further processing, the cheese should be wrapped or placed in an air-tight container and left to age at 35° F to develop flavor.

VII. Review Questions

A. Would Benedict's give a positive reaction with all carbohydrates? _____

Why? _____

_____

B. Put a drop of iodine on the corner of this page. Does paper contain starch? _____

C. What chemical elements are present in starch? _____

D. What is a reducing sugar? _____

E. What are the component molecules making up triglycerides? _____

F. Dry cleaners often use carbon tetrachloride. What stains would this remove? _____

_____

G. What are the building blocks of proteins? _____

H. How would you determine whether milk contains reducing sugars? _____

_____

I   What could you say about a substance which did **not** turn blue-black with iodine? _____

_____

J. Is CO_2 (carbon dioxide) considered organic? _____ Why? _____

_____

K. What is soap used for? _____

L. What does "hydrolysis" mean? _____

M. The hydrolysis of sucrose resulted in the presence of what two sugars? _____

Were they reducing sugars? _____

N. What was Little Miss Muffet actually eating? _____

CELL DIVISION

I. Objectives

After the completion of this exercise, the student should be able to do each of the following:

A. List the events that occur as a plant cell divides.

B. List the events that occur as an animal cell divides.

C. Answer the review questions at the end of this exercise.

II. Introduction

The first part of the cell theory states that cells are the structural and functional units of living organisms.

The second part of the cell theory states that all cells come from pre-existing cells. Cell division is the most common way of accomplishing this, although specific reproductive cells in most eukaryotic organisms are capable of another type of division process (meiosis). Cell division consists of two processes: **mitosis,** or nuclear division, and **cytokinesis,** or division of the cytoplasm and associated parts.

Cell division produces two daughter cells which are identical to the parent cell. This enabled you to grow from a fertilized egg into an adult human being; it is also the method by which cells are made to replace those killed or removed by a wound. In multicellular organisms, some cells give up the ability to divide in order to become specialized for some specific task. Except for such very specialized cells, the cells of any organism are potentially able to undergo division at any time, anywhere in the body. Most cells, however, divide only on special cues; cancer cells are cells not obeying the cues, but dividing wildly.

In brief contrast, meiosis can occur in only certain cells, at certain places in the body, at a certain time of the organism's life, and only in 2N cells. By this process the chromosome number (2N) is reduced to 1N.

Plant Cell Division

Since the chromosomes of plants are usually larger than those of animals and more easily seen, we will start with them, and go on to animal cells once the stages are identified here. Select a prepared, stained slide of *Allium* (onion) root tip.

The rounded end of the root is the tip, and is covered by the cells of the **root cap.** These cells have relatively thin cytoplasm and are not dividing. Their function is simply protection of deeper tissues. The cells which will be dividing are concentrated just behind this cap in an area which is called the **meristem.** The cytoplasm of these cells is much denser and many division figures should be visible if you move the slide around a bit. The cells are not dividing in sequence, so consecutive stages will not be right next to each other. Shop around until you find **all** of the stages listed.

Be able to identify any of the stages if shown under the microscope. (Just looking at the pictures is not enough. The real thing never looks like a diagram, and your test will be from the real thing, not the pictures.) Remember that these cells are dead, so they won't change stages before your eyes. Also any one stage can be subdivided into an early, middle and late stage, each of which will look just a tiny bit different.

Animal Cell Division

Pick out a slide of whitefish blastula. This is an embryonic stage of fish development. At the same time the embryo was sliced, it was essentially a ball of rapidly dividing cells. These embryos contain some oil as stored food for the embryo, and you will find dark purple spheres of this oil scattered over and among the individual cells.

Since virtually all of the cells are dividing, interphase will probably be the hardest to find. If a cell looks as though it has no nucleus, that is because the slice didn't happen to hit the nuclear material. The chromosomes are very tiny compared to those of plants, with the **asters** filling almost the whole cell.

Make sure that you can identify the stages diagramed if shown the real thing on a slide, as you did with plant cells.

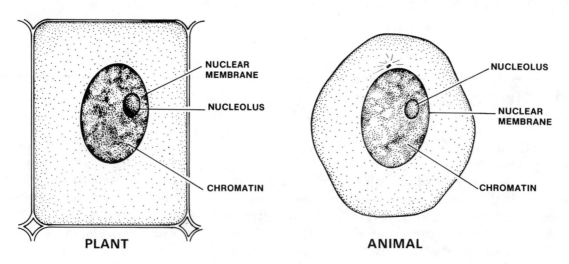

FIGURE 4.1
INTERPHASE

A. **Interphase**

1. DNA, present in a granular-appearing mass known as **chromatin,** replicates.

2. **Nucleolus** (i) is present.

3. **Nuclear membrane** is clearly visible.

36

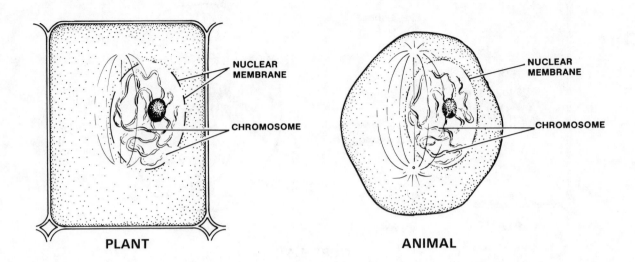

FIGURE 4.2
PROPHASE

B. **Prophase**

1. DNA coils tightly to form visible **chromosomes** which appear rod-shaped. One strand of a double-stranded chromosome is known as a **chromatid.**

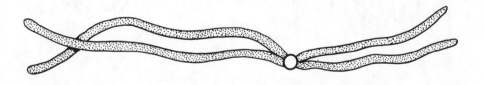

FIGURE 4.3
DOUBLE-STRANDED CHROMOSOME

2. Nuclear membrane dissolves.

3. Nucleolus (i) disappears.

4. In animal cells, a pair of **centrioles** (not visible on these slides) separate and with their newly formed **asters** (visible) move to each pole.

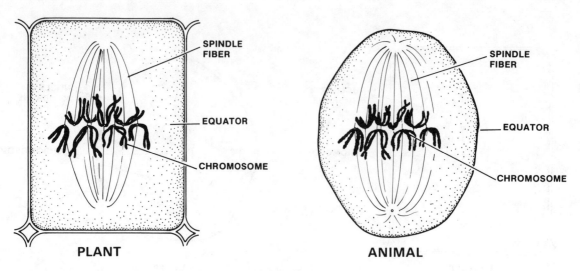

FIGURE 4.4
METAPHASE

C. **Metaphase**

    1.   Chromosomes line up at the **equator** of the cell.

    2.   One **spindle fiber** of the very conspicuous spindle system attaches to the **centromere** (**kinetochore**) of each chromosome.

FIGURE 4.5
ANAPHASE

D. **Anaphase**

    1.   Centromeres divide.

    2.   Single-stranded chromosomes (now consisting of one DNA molecule each) move to opposite poles.

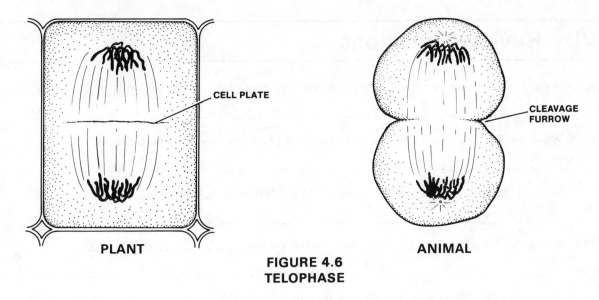

CELL PLATE

CLEAVAGE FURROW

PLANT

ANIMAL

FIGURE 4.6
TELOPHASE

E. **Telophase**

1. Chromosomes begin to uncoil into chromatin.

2. Nucleolus (i) and nuclear membrane begin to reappear.

3. Spindle slowly dissolves.

4. In animal cells, the centrioles double in number and the aster dissolves.

5. In animal cells, cytokinesis occurs with a **cleavage furrow** dividing one cell into two cells.

6. In plant cells, cytokinesis occurs with the formation of a **cell plate.** This plate divides the one cell into two cells and will later form the **middle lamella,** a common partition between adjacent cells.

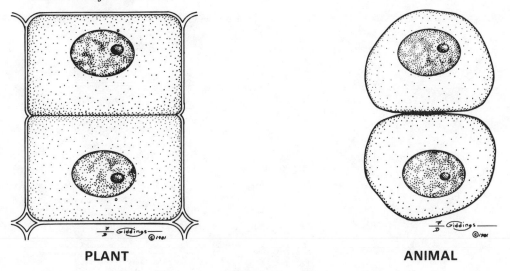

PLANT

ANIMAL

FIGURE 4.7
INTERPHASE (DAUGHTER CELLS)

F. **Daughter Cells**

This is the same as interphase, but these cells are smaller and less mature than the original cell.

VI.   Review Questions

A.   What are the most obvious features of interphase in plant or animal cells? _Ydar_
nuclear membrane, Chromatin, + nucleus

B.   What are the two most obvious differences in plant and animal cell division? _Mitosis_
_____

C.   What is the difference between mitosis and cytokinesis? _Mitosis is nuclear_
division. Cytokinesis is division of cytoplasm _part_

D.   Name the 4 major phases of mitosis in correct order _Interphase,_
Prophase, Metaphase, Anaphase, Telophase

E.   What process is busily occurring in a meristem? _Cell division_
_____

ENERGY CAPTURE AND CONVERSION BY THE CELL

exercise 5

I. Objectives

After the completion of this exercise, the student should be able to do each of the following:

A. Explain the function of mitochondria and recognize them in specially-prepared slides.

B. Write the generalized formulas for respiration and photosynthesis.

C. Describe a simple test for CO_2.

D. Explain the results of the experiment involving the respirometer.

E. Explain the experiments which involved chromatography, the relationship between duration of exposure of leaves to light and starch formation, and CO_2 consumption by a photosynthesizing plant.

F. Answer the review questions at the end of this exercise.

II. Respiration

A. Mitochondria

The mitochondria are the powerhouses of the cell; it is here that most of the organic fuels such as glucose are burned and ATP (an energy storage and transfer molecule) is produced. Although the chemical reactions of the process are intricate and extensive, the process may be briefly described by the following equation:

$$C_6H_{12}O_6 \text{ (glucose)} + O_2 \longrightarrow 6H_2O + 6CO_2 + \text{energy (to form ATP)}$$

On demonstration is a slide of tissue which was stained specifically to show mitochondria. Look for the small green-blue specks in the cytoplasm.

B. Carbon Dioxide Production

The CO_2 you exhale from your lungs is one of the by-products from burning or respiring fuel by the cells of your body. To prove that you exhale CO_2, use a straw to blow bubbles gently into a test tube of limewater. The CO_2 combines with the $Ca(OH)_2$ or **limewater** to form a white precipitate often known as **chalk.**

$$Ca(OH)_2 + CO_2 \longrightarrow CaCO_3 + H_2O$$

C. Oxygen Consumption

Seeds are essentially dormant plant embryos, accompanied by a carbohydrate fuel supply to help them until they can make leaves and roots of their own and make their own food. When soaked in water, the embryos come out of their dormancy and begin to respire their food supply. We will use germinating seeds to demonstrate oxygen consumption by putting them in an airtight system called a **respirometer.** The consumption of O_2 will cause a weak vacuum in the system.

However, if only the seeds were enclosed, their CO_2 production would cancel out the vacuum caused by O_2 disappearance. For this reason potassium hydroxide (KOH) is added to absorb the CO_2. Thus, only oxygen volume will change.

$$2\ KOH + CO_2 \longrightarrow K_2CO_3 + H_2O$$

We will also test seeds killed by soaking in formaldehyde (if they are completely dead, they shouldn't respire) and aquarium gravel (non-living things cannot respire). The gravel will also allow us to make adjustments during the test period if air pressure or room temperature should change. If the tube containing gravel should show changes in readings from either of these causes, we should have to add or subtract that amount from our live-seed data to be sure exactly how much gas was used. Such a setup is called a **control,** to make sure our results were really due to the experiment performed and not to some outside factor.

Each table should set up one live-seed and one dead-seed respirometer. **One** table in the class should also prepare a gravel control.

1. Half fill the tube with seeds (or gravel), add a wad of cotton, and then 5 or 6 pellets of KOH. (The cotton keeps the KOH from touching and killing the live seeds.) CAUTION: DO NOT HANDLE KOH WITH BARE FINGERS — IT IS LYE!

2. Insert the stoppers with their tubing and place a clamp on the vertical tube.

3. Adjust the respirometers in the ring stand clamps so the horizontal tubes are level.

4. Let apparatus sit for 2-3 minutes. The warmth of your hands on the tubes must dissipate, and the gas pressure must return to room pressure.

5. With a dropper add just enough dye to the horizontal tube to block the opening. As O_2 is used, the drop will be pulled past the ruler and you can make readings.

6. Make readings as indicated on the chart. (If the dye reaches the end of the ruler before readings are finished, unclamp the vertical tube and **gently** blow the dye back to the other end.)

7. CAUTION: in reading the tubes, don't stand too close. Body temperature can raise the temperature of the tubes and change the position of the drop.

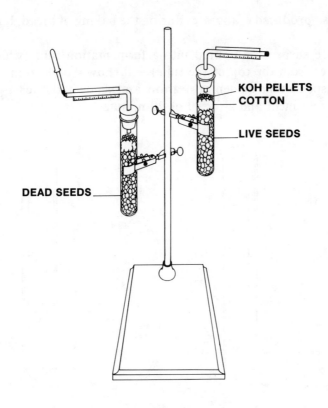

FIGURE 5.1
RESPIROMETER

TIME (MIN)	GRAVEL	LIVE SEEDS	DEAD SEEDS
		(CM TRAVELLED BY DROP FROM 1ST POSITION)	
0			
1			
2			
3			
4			
5			
6			
7			
8			
9			
TOTAL MOVED			

D. **Fermentation**

A type of fungus called common baker's yeast produces CO_2 and ethyl alcohol when respiring glucose. This doesn't require oxygen and is called **fermentation.** CO_2 bubbles make the bread rise. Other yeasts produce the CO_2 bubbles in beer and some wines. The ethyl alcohol which is also produced evaporates during the baking of bread, but is retained in beer and wine.

A yeast-glucose suspension poured into a fermentation tube (which is then tilted to remove any bubbles from the top of the tube) will show a collection of bubbles in a few minutes. Examine such a tube on demonstration, and shade in the yeast-glucose levels at the beginning of the experiment and after 10 or 20 minutes.

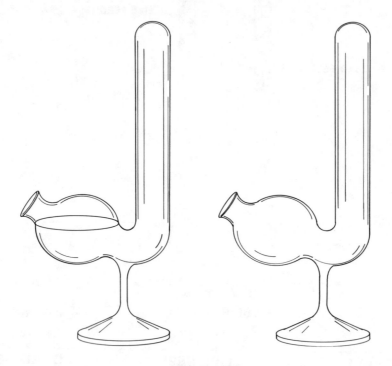

FIGURE 5.2
YEAST FERMENTATION

III.   Photosynthesis

The maintenance of life requires the continual use of energy. Radiant energy from the sun —this planet's only energy source — is captured by the process of photosynthesis and transformed into usable chemical fuels. All living things are directly or indirectly dependent upon this process.

The overall chemical equation for photosynthesis may be stated as follows:

$$6CO_2 + 6H_2O \quad \xrightarrow[\text{chlorophyll}]{\text{light energy}} \quad C_6H_{12}O_6 + 6O_2$$

A. CO_2 Consumption in Photosynthesis

1. Into a test tube which is two-thirds filled with water, place a large leafy stem of *Elodea* and then 4 or 5 drops of 1% solution of phenol red. The drops of phenol red should turn the water a red to pink color. **Phenol red** is a pH — indicating compound which gives a red color in a basic or neutral solution (pH 7 - 14) but the color changes to yellow if the solution becomes acidic (pH 0 - 6.9).

2. Now insert a straw into the test tube and blow gently **only** until the water becomes orange-yellow in color. Some of the exhaled CO_2 is dissolving in the water to form carbonic acid (popularly known as carbonated water).

$$H_2O + CO_2 \longleftrightarrow H_2CO_3$$

3. Then immediately cease the blowing, remove the straw, and stopper the tube. Place the test tube in a well-lit area for photosynthesis to resume. After 10-20 minutes note any color change.

   What was taken in by the *Elodea* plant during photosynthesis?_____

   _____

B. **Chromatography**

   Chloroplast and chromoplast pigment solutions are obtained by rupturing plant cells and extracting the pigments with an organic solvent such as acetone. (Caution: The solvents used in this exercise are quite flammable. Keep solutions stoppered to avoid excess inhalation or accumulation of fumes in the room.) Pigment extracts will be prepared already, by homogenizing a small quantity of leaves in a blender with acetone then filtering to remove cellular debris. This extract stains intensely, so avoid spilling.

   By a technique known as **paper chromatography,** mixtures of similar compounds may be separated on the basis of their different affinities for the solvent phase (acetone-ether) and the solid phase (paper) and hence different migration rates along the paper. Those with greatest affinity for the paper will travel slowest and those with greatest affinity for the solvent will travel the fastest.

1. Obtain a strip of filter paper which will fit easily into a test tube provided. Handle the paper only by the edges; oil from your fingers may interfere with the flow of the solvent.

2. Dip a micropipette into the extract. DO NOT cap the tube with your finger; the fluid will rise in the tube and remain by itself due to capillary action.

3. About 1 inch from the bottom of the paper, draw a quick line completely across, leaving as thin a line as possible.

4. Repeat pigment applications 3 or 4 times, drying between times to prevent a wide band from forming — blow or hold under a lamp.

5. Pour about ½ inch of chromatography solvent into the tube **first,** then insert the paper and stopper **loosely** with a cork (ether fumes expand in a warm room and may "pop" the cork out). The pigment line should not be submerged in the solvent.

6. Set the chromatogram aside for about 15 minutes to develop.

7. Remove the chromatogram before the uppermost color band runs off the top and allow to air dry.

8. Time and light cause the colors to fade out, so outline the top of each color band as soon as the chromatogram is dry. Then draw the color bands on the following chromatogram.

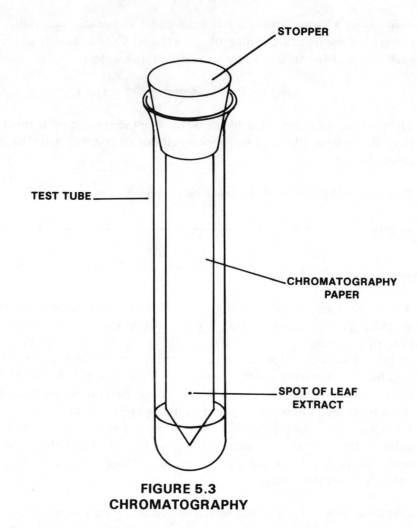

FIGURE 5.3
CHROMATOGRAPHY

9. You will note that a yellow band of **carotenes** moves about as fast as the solvent front. Lower on the paper will be one (or more) yellow **xanthophyll** bands, a bluish-green **chlorophyll a** band, a yellow-green **chlorophyll b** band, and in some chromatograms, a grey band between the carotene and xanthophylls; this is **pheophytin,** a breakdown product of chlorophyll indicating that the extract is becoming old. Red **anthocyanins** may show up at the base line as well. These are insoluble in the solvent. Label the bands on your chromatogram correctly.

C. **Duration of Light — Demonstration**

Starch, the carbohydrate storage product of photosynthesis, may be stored anywhere in the leaf where photosynthesis is occurring or be shipped to a root or leaf for storage. We will use *Coleus* leaves to show that only the leaf areas which contain chlorophyll can make and store starch. This plant has varicolored leaves; some areas contain only chlorophyll and look green, others contain the accessory pigment **anthocyanin** and look pink, still other areas contain no pigment and appear creamy white. Where chlorophyll and anthocyanin both occur, the color is a rusty brown-red.

The leaves were bleached in boiling alcohol, then stained with iodine to show the location of any starch deposits. Leaf A came from a plant which had been exposed to constant light for 72 hours prior to bleaching. Leaf B came from a plant which had been kept in the dark for 72 hours. Use colored pencils to indicate the areas of pigments and starch on the outlines following.

46

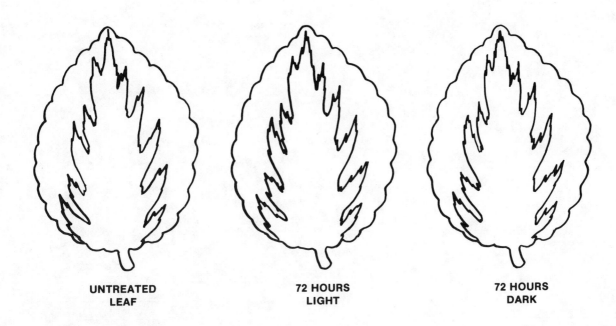

UNTREATED LEAF   **72 HOURS LIGHT**   **72 HOURS DARK**

FIGURE 5.4
COLEUS LEAVES

IV. Review Questions

A. Calcium carbonate ($CaCO_3$) is commonly known as _____ .

B. Are bean seeds living? _____ How can you tell? _____

_____

C. Name three products of fermentation by yeast. _____

D. Glucose is produced in which organelle? _____

E. Phenol red dropped into distilled water would produce what color? _____

F. What acid is produced by bubbling CO_2 into H_2O? _____

G. Name the bluish-green pigment in leaves. _____

H. Can an albino corn leaf produce any food? _____ Explain _____

_____

I. What is the main product of respiration? _____

of photosynthesis? _____

LIFE CLASSIFICATIONS AND REPRODUCTION

exercise 6

I. Objectives

After the completion of this exercise, the student should be able to do each of the following:

A. Know the various **taxa** used in classifying organisms.

B. List the five **kingdoms** of life.

C. Explain the difference between **asexual** and **sexual reproduction.**

D. Define the types of sexual reproduction.

E. Explain **alternation of generations** using appropriate terminology.

F. Answer the review questions at the end of this exercise.

II. Classifications

We will be considering diversity at various levels in the Monera, Fungi, and Plant kingdoms (and later in 102, kingdoms Protista and Animalia). You should, therefore, be familiar with the categories (called taxa) of this hierarchy. They are:

Kingdom
Phylum
Class
Order
Family
Genus
Species

Such a classification system is essentially an outline, much like one you might use to help you study any subject. Such systems could be based on almost any characteristics of the organisms involved, but scientists have chosen to use apparent common ancestry as their basis. Even so, many scientists differ in their analyses of common ancestry, so you will find that there are places in which minor differences appear when comparing different listings.

The following list gives the basic categories which we will study. This is by no means a complete list of all the categories which exist; it includes only those groups which we will examine or discuss as representative.

Kingdom Monera (moe-nair'-ah)
  Phylum Schizophyta (skits-ah'-fi-tah) - bacteria
  Phylum Cyanophyta (sy-ah-nah'-fi-tah) - blue-green algae

Kingdom Fungi (fun'-ji)
  Phylum Eumycophyta (you'-my-coff'-fi-tah) - true fungi
        Class Zygomycetes (zie'-go-my-see'-teez) ex: black bread mold
        Class Ascomycetes (ass'-koe-my-see'-teez) - sac fungi.
          ex: yeasts, morels, and *Aspergillus*
        Class Basidiomycetes (buh-sid'-ee-oh-my-see'-teez) - club fungi.
          ex: mushrooms and shelf fungi

Kingdom Plantae (plan'-tee)
  Subkingdom Thallophyta (thall-ah'-fi-tah)
        Phylum Rhodophyta (row-doff'-fi-tah) - red algae
        Phylum Phaeophyta (fay-ah'-fi-tah) - brown algae.
          ex: *Laminaria, Sargassum* and *Fucus*
        Phylum Chlorophyta (klor-ah'-fi-tah) - green algae.
          ex: *Spirogyra, Volvox,* and *Ulva*
  Subkingdom Embryophyta (em-bree-ah'-fi-tah)
        Phylum Bryophyta (bry-ah'-fi-tah)
          Class Hepaticae (he-pat'-ih-see) - liverworts
          Class Musci (moo'-see) - mosses
        Phylum Tracheophyta (tray-key-ah'-fi-tah) - vascular plants
          Class Filicineae (fill-ih-sin'-ee-ee) - ferns
          Class Gymnospermae (jim-no-sperm'-ee) - ex: *Pinus*
          Class Angiospermae (an-gee-oh-sperm'-ee) - flowering plants
              Subclass Dicotyledonae (di'-cotty-lee'-dun-ee) - ex: rosebush
              Subclass Monocotyledonae (mon'-oh-cotty-lee'-dun-ee) - ex: grass

Kingdom Protista (pro-tiss'-tah)
  Phylum Pyrrophyta (pye-rof'-fi-tah) - dinoflagellates. ex: *Gymnodinium,*
        Gonyaulax, and *Noctiluca*
  Phylum Chrysophyta (kriss-ah'-fi-tah) - golden algae. ex: diatoms
  Phylum Euglenophyta (you'-glen-ah'-fi-tah) - ex: *Euglena*
  Phylum Sarcodina (sar'-coh-die'-nah) - ex: *Amoeba,* radiolaria, and foraminifera
  Phylum Zoomastigina (zoe'-mas-tih-ji'-nah) - flagellates. ex: *Trypanosoma*
  Phylum Ciliophora (sill'-ee-ah'-for-ah) - ciliates. ex: *Paramecium*
  Phylum Sporozoa (spore'-oh-zoe'-ah) - ex: *Plasmodium* sp.

Kingdom Animalia (an-ih-male'-ee-ah)
  Phylum Porifera (pore-if'-er-ah) - sponges
        Class Calcarea (kal-care'-ee-ah) - chalky sponges.
          ex: *Grantia*
        Class Hexactinellida (hex'-act-in-el'-ih-dah) - glass sponges.
          ex: Venus flower basket
        Class Demospongiae (dem'-oh-spunge'-ee-ee) - commercial, or bath
          sponges. ex: *Spongia*

Phylum Cnidaria (nid-air'-ee-ah)
    Class Hydrozoa (hi'-droh-zoe'-ah) - hydroids. ex: *Hydra* and *Obelia*
    Class Scyphozoa (skife'-oh-zoe'-ah) - jellyfish. ex: *Aurelia*
    Class Anthozoa (an'-thow-zoe'-ah) - ex: sea anemones and corals

Phylum Ctenophora (ten-ah'-for-ah) - ex: comb jellies and sea walnuts

Phylum Platyhelminthes (plat'-ee-hell-min'-thees) - flatworms.
    Class Turbellaria (ter'-bel-air'-ee-ah) - free-living flatworms. ex: planaria
    Class Trematoda (treh'-mah-tode'-ah) - flukes. ex: *Clonorchis*
        and sheep liver fluke
    Class Cestoda (sess-tode'-ah) - tapeworms. ex: *Taenia* sp. and
        Dipylidium caninum

Phylum Nematoda (nee'-mah-tode'-ah) - roundworms. ex: *Ascaris,* hookworms, and
    Trichinella sp.

Phylum Rotifera (row-tiff'-fer-ah) - rotifers, or "wheel animals"

Phylum Annelida (an-nell'-ih-dah) - segmented worms.
    Class Archianellida (ark'-ee-an-nell'-ih-dah) - primitive marine worms
    Class Polychaeta (pahl'-ee-keet'-ah) - ex: *Nereis,* bristleworms
    Class Hirudinea (hi'-roo-din'-ee-ah) - leeches
    Class Oligochaeta (oh'-lig-oh-keet'-ah) - ex: earthworm

Phylum Arthropoda (arr-throp'-oh-dah) - joint-footed animals
    Subphylum Trilobita (try'-low-bite'-ah) - trilobites (extinct)
    Subphylum Chelicerata (kell-liss'-er-ah'-tah)
        Class Merostomata (mare'-oh-stow-mah'-tah) - ex: *Limulus*
            (horseshoe crab)
        Class Arachnida (uh-rack'-nid-ah) - ex: spiders, mites, ticks,
            scorpions and harvestmen
    Subphylum Mandibulata (man-dib'-you-lah'-tah) -
        Class Crustacea (kruss-tay'-she-ah) - ex: shrimp, crabs, lobsters,
            crayfish, and barnacles
        Class Insecta (in-sect'-ah) - insects. ex: bees, flies, beetles,
            butterflies, and bugs
        Class Chilopoda (ky-lop'-oh-dah) - centipedes
        Class Diplopoda (dop-lop'-oh-dah) - millipedes

Phylum Mollusca (moll-lusk'-kah) - soft-bodied animals
    Class Amphineura (am'-fee-noor'-ah) - chitons
    Class Monoplacophora (mon'-oh-plak-ah'-for-ah) - ex: *Neopalina*
    Class Bivalvia (bi-val'-vee-ah) - bivalves. ex: clams, oysters, mussels,
        and scallops
    Class Gastropoda (gas-trop'-poh-dah) - ex: snails, slugs, whelks, and sea hares
    Class Cephalopoda (seff'-fal-lop'-oh-dah) - ex: squids, octopus, and
        chambered nautilus
    Class Scaphopoda (skaff-op'-oh-dah) - ex: tuskshells

Phylum Echinodermata (ee-ky'-no-derm-ah'-tah) - spiny skinned animals
    Class Echinoidea (ek'-in-oyd'-ee-ah) - sea urchins and sand dollars
    Class Holothuroidea (hole'-oh-thure-oyd'-ee-ah) - sea cucumbers
    Class Crinoidea (krin-oyd'-ee-ah) - sea lillies and feather stars
    Class Ophiuroidea (oh'-fee-you-roy'-dee-ah) - brittle stars and serpent stars
    Class Asteroidea (ass-ter-roy'-dee-ah) - starfish

Phylum Hemichordata (hem'-ee-kord-ah'-tah) - acorn worms. ex: *Dolicho-*
    glossus and *Balanoglossus*

Phylum Chordata (kord-ah'-tah)
    Subphylum Urochordata (you'-row-kord-ah'-tah) - sea squirts, tunicates.
        ex: *Molgula*
    Subphylum Vertebrata (vur'-tih-brah'-tah) - vertebrates
        Class Agnatha (ag-nath'-ah) - jawless fish. ex: lampreys, hagfishes
        Class Placodermi (plak'-oh-der'-mee) - extinct armored fish with jaws
        Class Chondrichthyes (kon'-dra-ick'-thees) - cartilaginous fish.
           ex: sharks, skates, and rays
        Class Osteichthyes (oss'-tee-ick'-thees) - bony fish.
           ex: perch, seahorses, mackerel, pike
        Class Amphibia (am-fib'-ee-ah) - ex: salamanders, frogs, and toads
        Class Reptilia (rep-till'-ee-ah) - ex: turtles, snakes, lizards, and crocodiles
        Class Aves (ah'-vees) - birds. ex: chicken, sparrow, eagle
        Class Mammalia (mam-male'-ee-ah) - ex: duckbilled platypus,
           kangaroo, bat, rat, apes, man

III. Reproduction

Asexual reproduction is any method of producing new individuals without the fusion of **gametes,** or sex cells, and in which only one organism is involved. This is done routinely in very primitive plants, producing offspring identical to the parent. Often very elaborate structures are produced for this purpose. When no deliberate preparations are made by the parent for the production of new individuals, the process may be called **vegetative reproduction.** In some cases the center of a branched body dies and the branches left continue growth. Fragmentation may occur accidentally, of course, as when a grazing animal bites off a chunk and leaves scraps behind. Early land plants lost the ability to deliberately reproduce asexually as they became more advanced, but virtually all plants can "recreate" themselves if the scrap is large enough.

Unlike the above, **sexual reproduction** produces offspring potentially different from the parents. It involves the fusion or fertilization of two cells (or **gametes**) to produce a new cell called a zygote. Gametes may look identical (a condition known as **isogamy**) or be visibly different **(heterogamy).** The most common type of heterogamy is known as **oogamy,** in which large, non-motile **eggs** are fertilized by tiny, motile **sperm.**

Most organisms are either **haploid** (1N) or **diploid** (2N), but during reproduction, intermediate cells or resulting products may have a different chromosome number. This change of the chromosome number is created by nuclear fusion of gametes or by meiosis. The reproductive history of how an organism produces another organism similar to itself is known as its life cycle.

While there are thousands of life cycles found in nature, most organisms follow basically one of the following three schemes.

The life cycle diagrammed in Figure 6.1 is found predominantly in the Protista and Fungi Kingdoms as well as in the algae of the Plantae Kingdom. The major and most conspicuous organisms are haploid. They produce gametes mitotically or else their haploid nuclei act as gametes. Fertilization, or nuclear fusion, produces a **zygote** which is usually short-lived. The zygote is diploid, and it undergoes meiosis to produce 1N spores or just young haploid organisms.

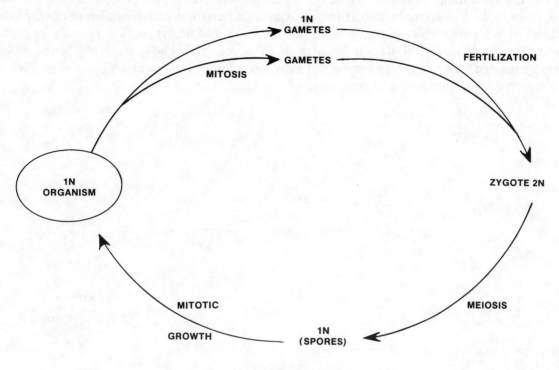

FIGURE 6.1
LIFE CYCLE COMMON IN FUNGI

The life cycle diagrammed in Figure 6.2 is found in all animals, some fungi and some algae. The conspicuous diploid organism produces gametes meiotically. These fuse to form a zygote which is diploid. The zygote then grows mitotically to become a multicellular diploid organism. The gametes are the only haploid stage in this type of life cycle.

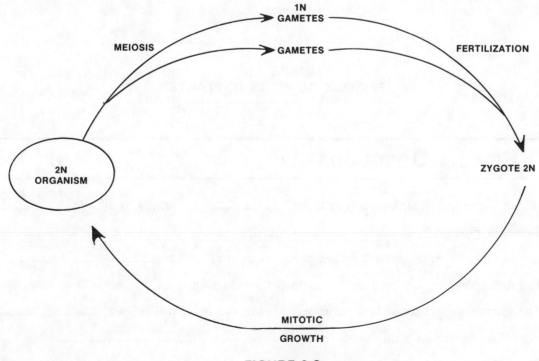

FIGURE 6.2
LIFE CYCLE COMMON TO ANIMALS

The life cycle diagrammed in Figure 6.3 is found exclusively in the Kingdom Plantae. Any one species exists in two multicellular forms. A multicellular haploid organism **(gametophyte)** produces gametes mitotically. These gametes fuse to form a diploid zygote. The zygote then grows mitotically to form a multicellular diploid organism **(sporophyte).** The sporophyte then produces **spores** by meiosis, and the spores grow mitotically into multicellular gametophytes. This cycle is often referred to as **alternation of generations.**

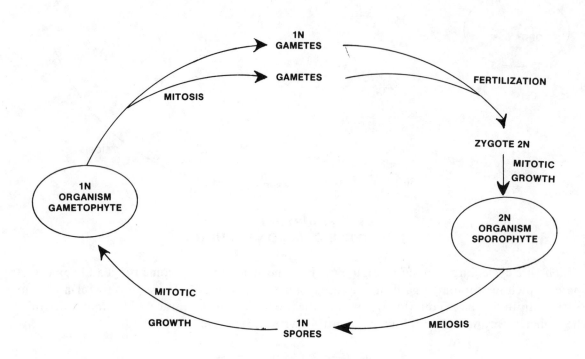

**FIGURE 6.3
LIFE CYCLE COMMON TO PLANTS**

IV. Review Questions

A.   Which is the higher taxon, class or order? _____ Family or class? _____

Genus or phylum? _____

B.   Is isogamy a form of asexual, or sexual reproduction? _____

C.   Egg and sperm are involved in: a. asexual reproduction   b. isogamy   c. oogamy

D.   The sporophyte plant has how many sets of chromosomes, one or two? _____

E.   Which generation in plants produces spores? _____

F.   Plant eggs and sperm, unlike animals, are produced by which process — mitosis or meiosis?

KINGDOM MONERA

I. Objectives

After the completion of this exercise, the student should be able to do each of the following:

A. Draw and identify **coccus, bacillus,** and **spirillum** bacterial cell shapes.

B. Make a Gram-stained smear of bacteria and identify their Gram reactions.

C. Know the identifying traits of *Oscillatoria* and *Nostoc*.

D. Know the meaning of **nitrogen fixation, trichome, sheath, heterocyst, colony,** and **fission.**

E. Explain the difference between a colonial and a multicellular organism.

F. Answer the review questions at the end of this exercise.

II. Introduction

Members of the kingdom Monera show a simpler type of cellular organization than do other living organisms. Their cells are **prokaryotic,** that is, they lack the complex membrane-bound organelles found in cells such as those you observed in exercise 1. The cells you observed in that exercise were **eukaryotic.** They possessed mitochondria, chloroplasts, Golgi bodies, etc. which prokaryotic cells lack. The few non membrane-bound organelles which prokaryotic cells *do* have are different in size and/or structure from those of eukaryotic cells (ex. ribosomes, flagella, etc.)

The kingdom Monera is divided into phyla on the basis of nutritional modes and types of photosynthetic pigments present, if any.

III. Phylum Schizophyta — bacteria

Bacteria can be **heterotrophic,** deriving their energy from organic molecules made by other living organisms, or **autotrophic,** deriving the energy they need from photosynthesis or from the oxidation of inorganic molecules. Most are heterotrophs, but those which are capable of photosynthesis differ from members of the other phyla in lacking chlorophyll a.

Bacteria are mostly unicellular or form chain-like or clustered groupings of cells. Their reproduction is primarily by a simple asexual process of cell division known as **fission,** in which a cell pinches into two without the complex movement of chromosomes seen in mitosis. Several bacterial species also engage in a type of sexual reproduction known as **conjugation,** in which all or part of the DNA of one bacterium is transferred to a second bacterium through a specialized cell projection known as an F pilus.

Bacteria usually have cell walls which frequently contain components of chitin - a substance also making up the exoskeletons of insects and their relatives. It is the cell wall which gives bacteria their characteristic shapes. Among the largest class of bacteria, the **Eubacteria,** there are only three basic cell shapes:

spiral - **spirillum**
spherical - **coccus**
rod - **bacillus**

Spirilla are always single cells, but cocci and bacilli may show a variety of cell arrangements because their cells tend to stick together after fission. Both bacilli and cocci may be paired or in chains **(filaments),** and cocci may also be clustered in various ways. Cell arrangement is characteristic for each species of bacteria. These loose arrangements of cells which are basically independent functional units are called **colonies.** Colonial cell arrangements differ from the multicellular state seen in higher organisms in that the cells making up multicellular organisms are dependent for survival on one another to varying degrees.

Besides the Eubacteria, there are five other bacterial classes:

Rickettsias and **Mycoplasms** are so small that they challenge the resolving power of the compound light microscope.

Myxobacteria lack cell walls and move about like tiny amoebas.

Actinomycota grow in branching chains and produce many of the antiobiotics in use today.

Spirochetes, which form long, thin, flexible spirals, are best known as the group to which *Treponema pallidum,* the cause of syphilis, belongs.

A.   The demonstration microscopes show a variety of bacteria of each of the cell shapes and arrangements mentioned. Make sketches of bacteria showing each of the main cell shapes.

B.   The Gram Stain

One of the most important procedures in identifying bacteria of medical importance is the **Gram staining technique.** Because of differences in cell wall composition, bacteria respond to this staining technique in one of two different ways, either retaining the initial violet stain when rinsed with alcohol, or bleaching out. Those which retain the stain are said to be **Gram positive.** Those which bleach out are said to be **Gram negative.**

This difference in staining properties corresponds to differences in susceptibility to different antiobiotics, so it is often possible to begin treatment of an infection before the exact identity of the bacteria causing it is known, a process which requires several days at least.

Few bacteria cause infection. There are hundreds of species of bacteria living in or on the human body, including the oral cavity, which are nonpathogenic (not disease-causing).

In this activity, you will prepare a smear from scrapings taken with a toothpick from your teeth and perform the Gram staining technique on it. Try to find at least two of the three cell shapes and bacteria showing both Gram positive and Gram negative staining properties.

1.   Use the wide end of a toothpick to scrape your teeth near the gum line, then mix the scrapings thoroughly in a small drop of water at one end of a clean slide.

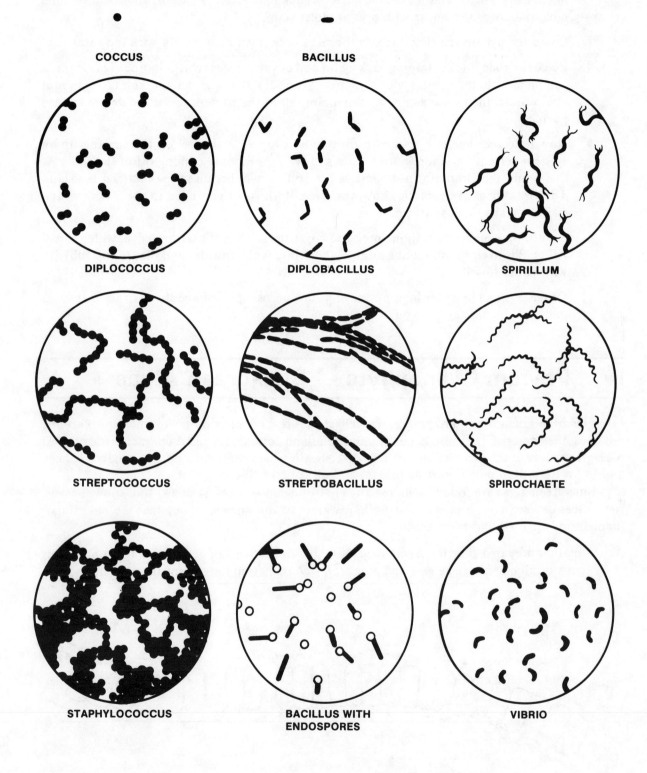

COCCUS

BACILLUS

DIPLOCOCCUS

DIPLOBACILLUS

SPIRILLUM

STREPTOCOCCUS

STREPTOBACILLUS

SPIROCHAETE

STAPHYLOCOCCUS

BACILLUS WITH
ENDOSPORES

VIBRIO

FIGURE 7.1
BACTERIAL SHAPES AND ARRANGEMENTS

2. Allow the smear to dry completely, then heat it slightly by holding it briefly over your microscope's light. This makes the bacteria stick to the slide. Place the slide on a staining rack and cover the smear with crystal violet stain.

3. Leave the dye on the slide for about one minute, then rinse gently with tapwater.

4. Place the slide on the staining rack again and cover the smear with iodine. This forms a complex with the crystal violet. Allow the iodine to stay on the smear for about one minute, then rinse again. At this point, all of the bacteria on the slide are stained purple.

5. Now hold the slide in a slanted position and drop 95% alcohol on the smear with an eyedropper until no more purple stain shows in the alcohol coming off of the slide. At this point, the Gram positive bacteria are still purple, because they retained the stain. The Gram negative bacteria, however, are colorless, so you have to use a contrasting stain to make them show up.

6. Return the slide to the staining rack and cover the smear with safranin dye, which is red. Allow this to sit for about one minute, then rinse with tapwater and blot (don't rub) dry with paper towelling.

7. Observe the slide under high power of your microscope or have the instructor focus it under oil immersion for you.

IV. Phylum Cyanophyta — Blue-green Algae

The blue-green algae are similar in cell structure and physiology to their prokaryotic relatives, the bacteria. They also occur in unicellular and colonial forms. A characteristic of most blue-green algae is the production of a jelly-like **sheath** which surrounds both the individual cell and the colony. Reproduction is by fission, and all contain chlorophyll a.

Blue-green algae are found in almost any environment — lakes, streams, soil, snow, oceans, rocks, ice, deserts, and even the near-boiling waters of hot springs. They may be unicellular, filamentous, or form sheets of cells.

A. Obtain a prepared slide of *Oscillatoria*. Note the long chains of cells called **trichomes.** This genus of blue-green algae gets its name from its oscillating movement through the water.

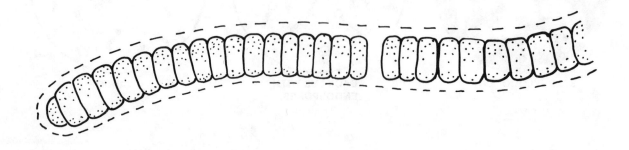

FIGURE 7.2
OSCILLATORIA

B. Prepare a slide of preserved *Nostoc. Nostoc,* which was once referred to as "witches' butter" or "starjelly", forms large, grape-like microscopic "colonies." The microscopic colonies contain two types of cells in their trichomes. The smaller are vegetative cells; the larger are known as **heterocysts,** in which **nitrogen fixation** is believed to occur.

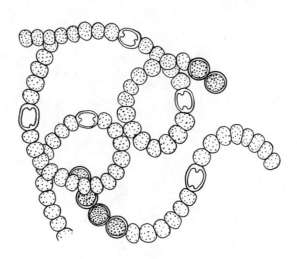

FIGURE 7.3
NOSTOC

V. Review Questions

A. Which organism is autotrophic: *Nostoc* or a spirochete? _____

B. What is meant by the term "gram positive"? _____

C. Why does *Oscillatoria* deserve its name? _____

D. What are the three basic types of bacterial shapes? _____

_____

E. What is nitrogen fixation? _____

_____

F. What important characteristics do bacteria and blue-green algae have in common? _____

_____

_____

G. Tell the main difference between blue-green algae and bacteria. _____

_____

_____

KINGDOM FUNGI (PHYLUM EUMYCOPHYTA)

exercise 8

I. Objectives

After the completion of this exercise, the student should be able to do each of the following:

A. List the major characteristics of fungi.

B. Describe the typical body of a fungus, or **mycelium.**

C. Identify *Rhizopus* with its four types of hyphae **(rhizoids, stolons, sporangiophores,** and **gametangia)** and explain its life cycle.

D. Explain the differences between the Zygomycetes, Ascomycetes, and Basidiomycetes and distinguish fresh and preserved specimens of each group.

E. Distinguish between pore, cup, gill, and tooth fungi.

F. Explain **symbiosis,** list the three major types, and give an example of each.

G. Tell what a **lichen** consists of and how it reproduces. Identify the three basic forms of lichens.

H. Answer the review questions at the end of this exercise.

II. Characteristics of the Fungi

The fungi are a large, diverse group of **eukaryotic** organisms. They possess cell walls but these cell walls are usually made of **chitin** instead of cellulose. Fungi lack chlorophyll so they feed themselves by being **saprophytes** (which live off dead organisms) or **parasites** (which live off living organisms). Thus, the fungi are valuable agents of decay for recycling simple compounds and elements needed for new life.

The **mycelium** or body of a fungus is usually composed of a mass of branching filaments called **hyphae.** For the most part, the mycelium is seldom seen since it grows deep inside its host.

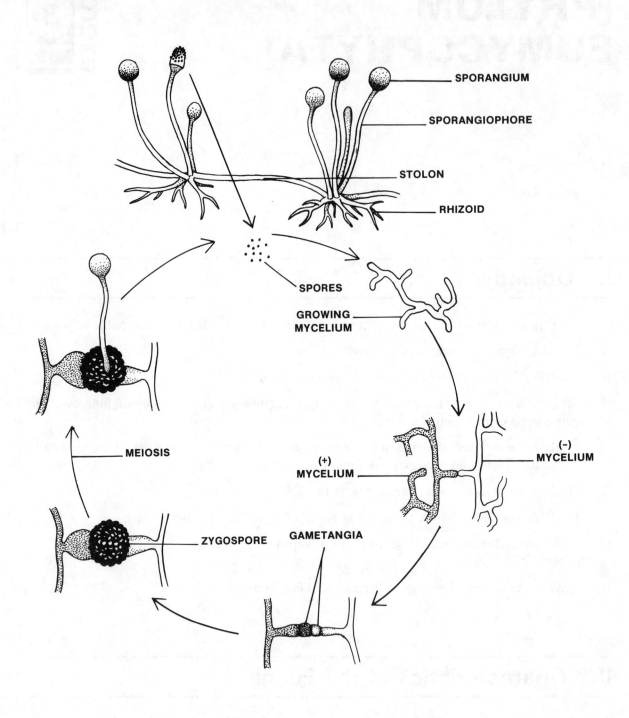

SPORANGIUM

SPORANGIOPHORE

STOLON

RHIZOID

SPORES

GROWING MYCELIUM

(+) MYCELIUM

(−) MYCELIUM

MEIOSIS

ZYGOSPORE

GAMETANGIA

FIGURE 8.1
RHIZOPUS LIFE CYCLE

The reproductive structures, which produce spores either by mitosis or meiosis, are usually visible and characteristic of the species.

Although both asexual and sexual reproduction are possible for most fungi, there are no morphological differences which could be designated male and female; therefore, the term "mating strains" is used and the types designated as plus (+) and minus (−). A plus strain can only "mate" with a minus strain.

III. Class Zygomycetes — Bread Molds

This is the most primitive of the three fungal classes we will examine. Zygomycete hyphae have no **cross walls (septa)** between the nuclei. Their mycelium is therefore a mass of communal cytoplasm, making separate compartments for nuclei only during reproduction.

Rhizopus has several basic types of hyphae: **rhizoids** for food absorption, **stolons** to enable the fungus to spread quickly over the substrate surface, and **sporangiophores** with **sporangia** for asexual reproduction.

In sexual reproduction, short lateral hyphae form **gametangia** which contain either (−) or (+) strain nuclei as the gametes. The gametangia fuse to form a zygospore. The zygospore contains numerous diploid (2N) nuclei and develops a dark rough wall. Inside this zygospore, one diploid nucleus undergoes meiosis to produce one viable 1N nucleus. This haploid organism grows and produces an asexual sporangiophore.

A. Obtain a prepared slide of *Rhizopus nigricans,* the common black bread mold. Look for rhizoids, stolons, sporangiophores, gametangia, sporangia and zygotes.

B. Observe *Rhizopus* growing on various substrates such as bread, beans, agar, fruit, etc. Notice the sporangiophores, stolons, and rhizoids.

IV. Class Ascomycetes — Sac Fungi

Ascomycetes are usually characterized by **septate** hyphae and by sexual reproductive structures called **asci** which contain eight **ascospores.** The asci are often found in large numbers within structures called **ascocarps** such as morels and cup fungi.

A. Make a fresh slide of common baker's yeast. Yeast have evolved to exist without hyphae and their main means of reproduction is asexual **budding.**

B. Examine various types of common sac fungi such as *Penicillium, Aspergillus,* morels, and cup fungi. The first three illustrations in Figure 8.2 show asexual structures. **Note** the illustrations are not drawn to scale.

V. Class Basidiomycetes — Club Fungi

This and the ascomycete class have a unique stage in their life cycles found in no other groups — the **dikaryon.** At gamete fusion in other forms, both the cytoplasms and nuclei fuse immediately. In the dikaryon, the cytoplasm of the gametes fuses but the nuclei remain separate

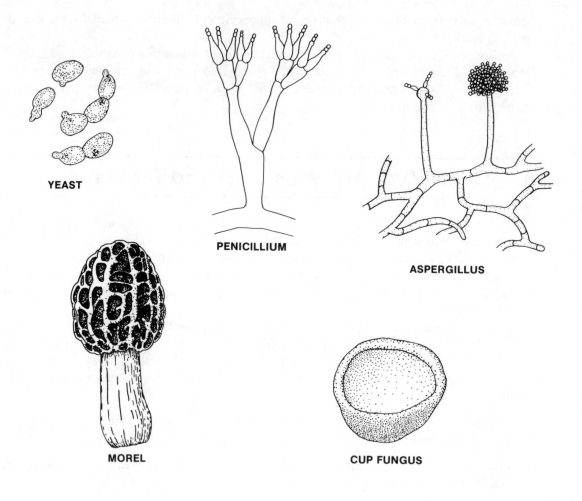

YEAST

PENICILLIUM

ASPERGILLUS

MOREL

CUP FUNGUS

FIGURE 8.2
VARIOUS ASCOMYCETES

and divide in tandem for a varying period of time. This produces a multicellular body which is not really 1N (since there are two sets of chromosomes in one cell) but not actually 2N either (since the two sets are not within one nuclear membrane). This stage is designated as $N + N$, or **dikaryotic.** In the ascus or basidium, just before meiosis, the two nuclei fuse to become a genuine 2N nucleus.

Basidiomycetes have septate hyphae and sexually produce spores externally on cells called **basidia.** Thousands of basidia occur in a fruiting body called a **basidiocarp.** The basidiocarp may be a mushroom, a puffball, or a slimy mass of gelatinous hyphae.

A. **Coprinus mushroom**

Examine a prepared slide of a x.s. of *Coprinus* basidiocarp. Note the basidia, the **sterigmata** which hold the spores to the basidia, the gills, and spores.

B. While some basidiocarps form basidia on long gills underneath the cap, others form basidia in deep narrow pores or on projecting teeth. Examine specimens of mushrooms and other fungi and note which have gills, pores, or teeth. Some fungi live in trees and produce woody basidiocarps. These are often called **shelf fungi.**

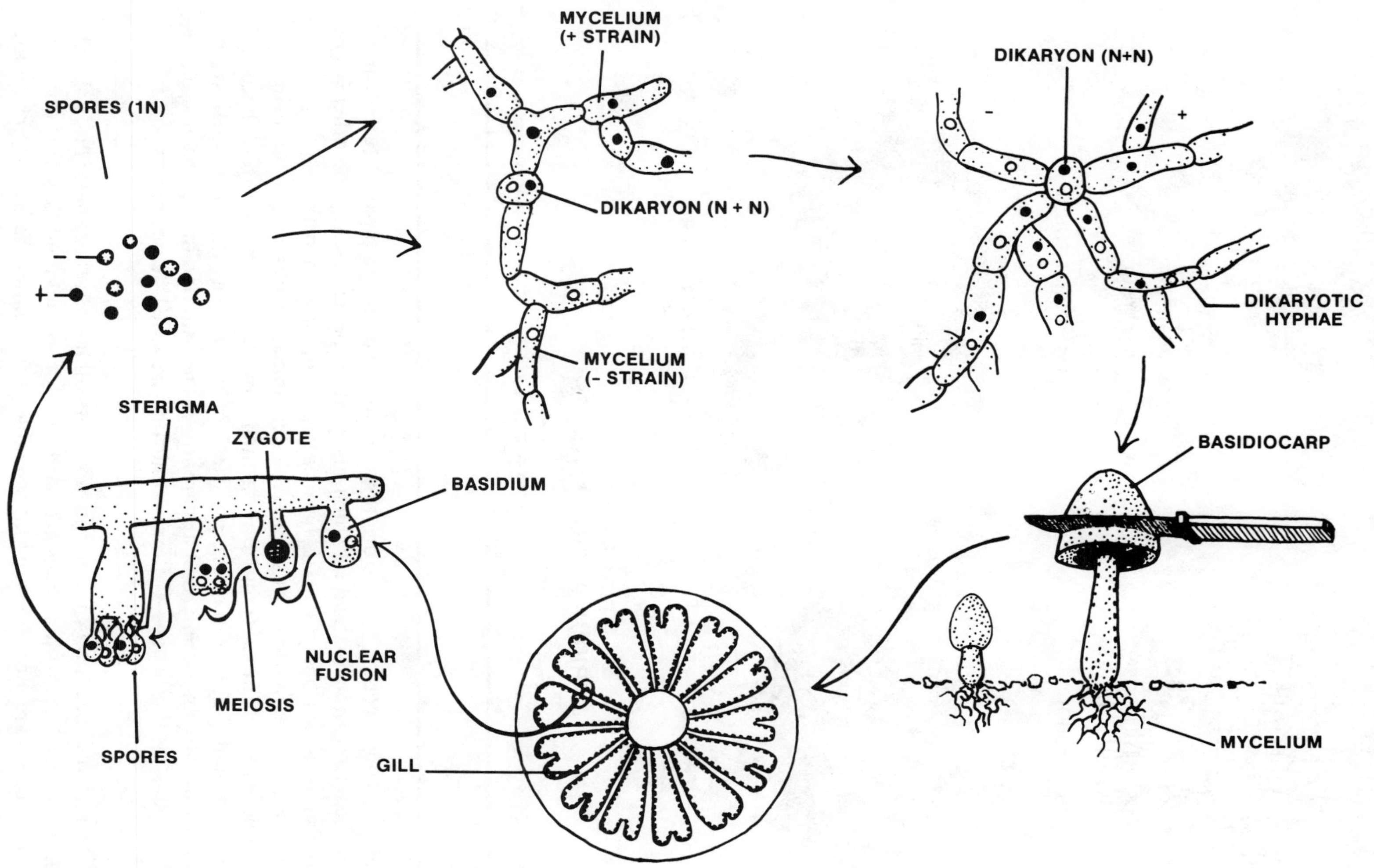

FIGURE 8.3
BASIDIOMYCETES LIFE CYCLE

DIKARYOTIC HYPHAE

DIKARYON (N+N)

BASIDIOCARP

MYCELIUM

MYCELIUM (+ STRAIN)

DIKARYON (N + N)

MYCELIUM (– STRAIN)

GILL

BASIDIUM

NUCLEAR FUSION

MEIOSIS

ZYGOTE

STERIGMA

SPORES

SPORES (1N)

SPORES

65

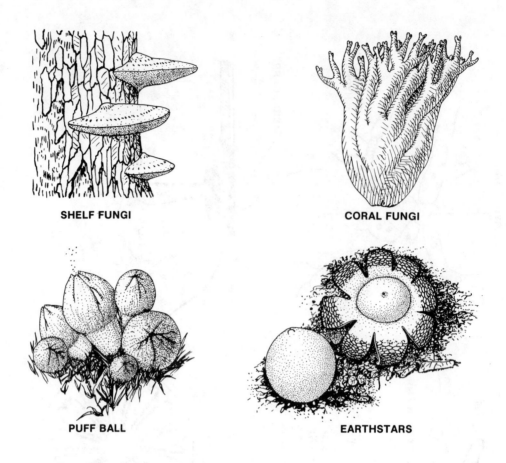

SHELF FUNGI

CORAL FUNGI

PUFF BALL

EARTHSTARS

**FIGURE 8.4
VARIOUS BASIDIOMYCETES**

VI.   Lichens

Symbiosis means "living together." There are three major types of symbiosis: **Mutualism** in which both partners benefit, **commensalism** in which one partner benefits while the other is not affected, and **parasitism** in which one partner is harmed while the other benefits.

Lichens are examples of the first type of symbiosis, a combination of an algae (contributing food from photogynthesis) and a fungus (contributing anchorage and retaining water). The algal components of lichens (usually blue-green or green algae) can be grown by themselves. However, the fungal component (usually an ascomycete) apparently cannot survive in the absence of the proper alga. For this reason, the lichens are taxonomically classified according to the fungal partner.

A.   We will not examine any specific taxonomic groups, but will look at three basic body shapes found among the lichens. **Crustose** lichens are hard, flat forms looking like stains on rocks or wood. **Foliose** lichens are more leaflike and resemble peeling paint. **Fruticose** lichens are usually erect, branching structures. Examine examples of these three types as available in the lab.

B.   Most of the algae and fungi of the lichens can reproduce independently, but the symbiotic unit reproduces by pinching off chunks of itself **(soredia)** containing a few algal cells and a few hyphae. On a prepared slide of a lichen, note the algal cells near the surface, fungal hyphae and soredia, if present.

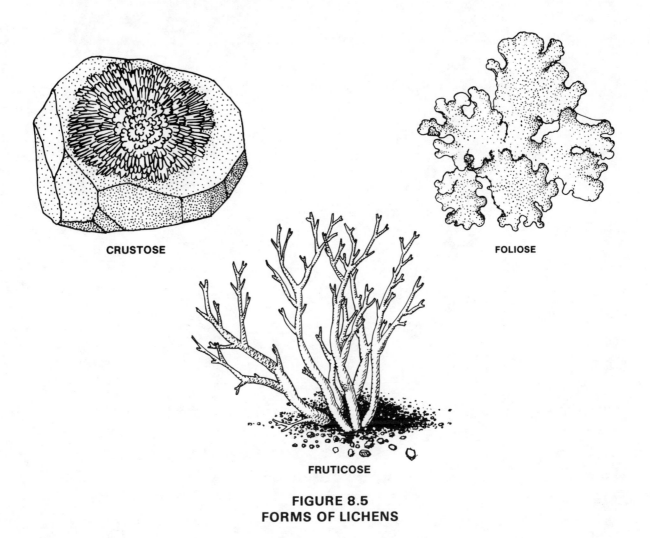

CRUSTOSE

FOLIOSE

FRUTICOSE

**FIGURE 8.5
FORMS OF LICHENS**

VII.   Review Questions

A.   A shelf fungus produces which structure: basidiocarp, ascocarp, or lichen? _____

B.   Saprophytes are: a. autotrophic   b. heterotrophic   c. parasitic

C.   *Rhizopus* reproduces sexually by _____ (isogamy; heterogamy).

D.   What economic importance do some ascomycetes have? _____

E.   If you pick a mushroom, do you kill the entire fungal body? _____ Why?

_____

F.   Budding is a form of _____ (sexual; asexual) reproduction.

G.   What nuclear arrangement is unique to the ascomycetes and basidiomycetes? _____

_____

KINGDOM PLANTAE — ALGAE

I. Objectives

After the completion of this exercise, the student should be able to do each of the following:

A. Discuss the basic characteristics of the green algae.

B. Describe the structures and/or reproductive mechanisms of *Protococcus*, *Spirogyra*, *Volvox*, and *Ulva*.

C. Discuss briefly the habitats and morphology of the red and brown algae.

D. Explain the sexual reproduction of *Spirogyra*.

E. Answer the review questions at the end of this exercise.

II. Phylum Chlorophyta — Green Algae

The green algae are the most diverse of all the algal groups. They range from single cells to colonies to multicellular organisms.

The cells of most green algae have a transparent cellulose wall and a prominent vacuole. The chlorophyll and carotene pigments are in approximately the same proportions as in most "higher plants." **Pyrenoids,** centers of starch formation, often occur on the chloroplasts.

Most green algae reproduce both asexually and sexually. Asexual reproduction occurs usually as mitotic cell division in unicellular and colonial forms, but multicellular organisms produce asexual spores and also fragment. Sexual reproduction is represented by both isogamy and heterogamy.

A. Unicellular Green Algae

Protococcus is one of the most common, and yet simplest, of the green algae. It can be found on tree bark, fence posts, flower pots, brick walls, damp soil and rocks. *Protococcus* reproduces only by cell division. Its spherical cell wall surrounds a large cup-shaped chloroplast.

1. Prepare a fresh mount of *Protococcus* and examine under high power.

2. Notice the example habitat for *Protococcus* such as tree bark or flower pots. Just like many mosses, Protococcus often grows more heavily on the moist, northern sides of tree trunks.

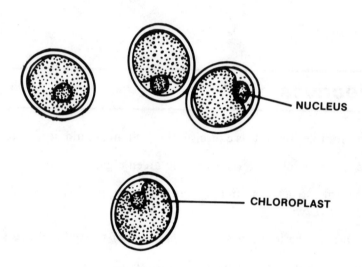

FIGURE 9.1
PROTOCOCCUS

B. **Colonial Green Algae**

1. One common form of colonial algae is **filamentous** in which cell division occurs in only one plane. The cylindrical cells of *Spirogyra* contain spirally-shaped chloroplasts with conspicuous pyrenoids.

Obtain a prepared slide of *Spirogyra*. Note the filaments with protrusions. Each is called a **conjugation tube,** through which the nucleus and the rest of the protoplasm of one cell act as an ameboid gamete and pass into the adjoining cell of the other filament. A **zygote** is produced which soon forms a thick wall and undergoes a resting stage **(zygospore).** Later, meiosis occurs to produce four haploid nuclei; three die. The remaining nucleus is incorporated into a new haploid cell from which a new filament develops by simple cell division. Label the illustration (Figure 9.2) with these terms: chloroplast, conjugation tube, nucleus, pyrenoid, and zygote.

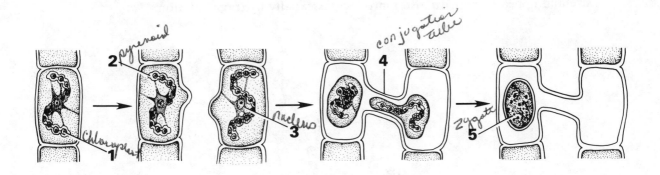

Labels on figure: 2 pyrenoid, conjugation tube, 4, nucleus, 3, chloroplast, 1, zygote, 5

FIGURE 9.2
SPIROGYRA

2. (OPTIONAL) Prepare a fresh mount of *Volvox*. *Volvox* are found in cool, fresh, calm waters. *Volvox* colonies vary in size from several hundred to as many as 40,000 cells arranged in the wall of a hollow sphere and interconnected by **plasmodesmata.** Asexual reproduction occurs through the formation of **daughter colonies** (see Figure 9.3) from single vegetative cells of the parent. These detach to the interior of the parent colony, and when the parent colony disintegrates from age, the daughter colonies swim away. Sexual reproduction is an advanced type of **heterogamy** (differing gametes) called **oogamy.**

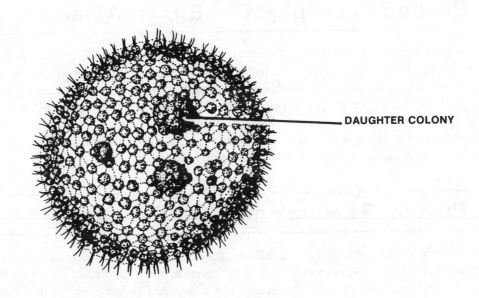

━ **DAUGHTER COLONY**

FIGURE 9.3
VOLVOX

C. Multicellular Green Algae

Ulva (sea lettuce) is one of the most commonly found green algae. *Ulva* clings to submerged rocks, oyster shells, and other substrates in salt or brackish waters. It undergoes alternation of generations in which the 2N sporophyte and 1N gametophyte are similar in size and appearance. *Ulva* also reproduces asexually by fragmentation.

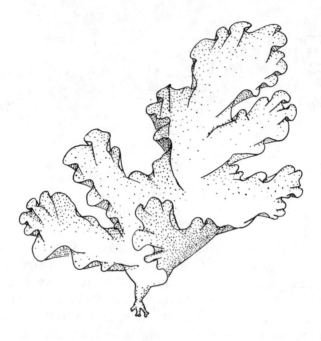

FIGURE 9.4
ULVA

III. Phylum Phaeophyta — Brown Algae

The brown algae are multicellular and almost exclusively marine. They include the largest of the seaweeds, the kelps, which may grow to 300 feet in length. A brown algae attaches to a solid surface by means of an adhesive **holdfast,** from which extends a stem-like **stipe,** bearing a flattened leaf-like **blade.** Deep species possess **air bladders** located along the blades, which float the plant's photosynthetic areas up into the lighted areas.

Observe various preserved specimens of brown algae. Note the holdfasts, stipes, blades, and air bladders. Label algae in Figure 9.5 with the previous terms.

IV. Phylum Rhodophyta — Red Algae

In addition to chlorophyll, the red algae contain two accessory photosynthetic pigments, **phycoerythrin** (red) and **phycocyanin** (blue). For this reason, they often are distinctly red in color. With these pigments, red algae can capture the rays which penetrate deepest into water, therefore can live deeper than most algae. This deep habitat is often below the levels of major currents, in nearly still water, enabling them to have very fragile bodies. These delicate and beautiful forms make them the darlings of seaweed collectors and their flavors the favorites of European and Oriental gourmets. Examine herbarium specimens of red algae, noting the body form.

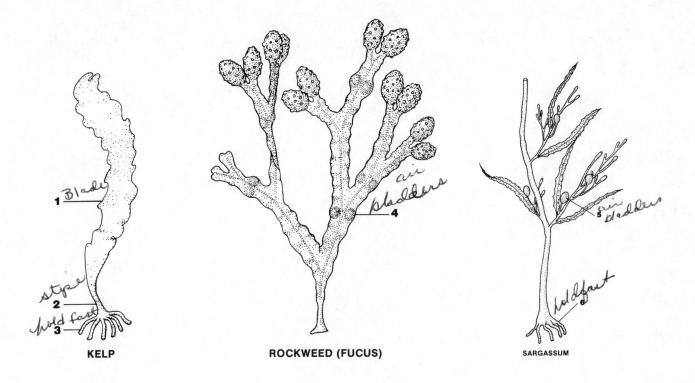

FIGURE 9.5
BROWN ALGAE

KELP

ROCKWEED (FUCUS)

SARGASSUM

V. Review Questions

A. What is the function of a pyrenoid? _Centers of starch formation_

B. Why does *Protococcus* usually grow more heavily on the northern sides of trees? ____
Because it is most moist

C. What is the function of air bladders? _float up to lighted areas_ Which phylum of
algae are they characteristic of? _Phaeophyta_

D. Explain these terms:

zygospore _resting stage when zygote is produced_
daughter colony _asexual reproduction, from single veg. cells of parent_
stipe _stem-like part of algae_
holdfast _adhesive like base_
filament _has protrusions_
conjugation tube _through which the nucleus + the rest of one
cell pass through into the adjoining filament_

E. What wavelengths of light would a red alga be able to capture which other green plants
could not capture? _They pass the penetrate the deeper_ Why? _Because of the
photosynthesis pigments phycoerythrin (red)
and phycocyanin (blue)_

73

KINGDOM PLANTAE —
PHYLUM BRYOPHYTA

exercise 10

I. Objectives

After the completion of this exercise, the student should be able to do each of the following:

A. Discuss the characteristics of the bryophytes.

B. Describe alternation of generations in liverworts such as *Marchantia,* and identify the associated structures.

C. Describe alternation of generations in a typical moss.

D. Answer the review questions at the end of this exercise.

II. Characteristics of Bryophytes

The phylum Bryophyta contains small, autotrophic organisms which have become adapted to a terrestrial existence. The members of this phylum exhibit the following characteristics:

A. They are green, have **rhizoids** (rootlike structures) and may have stemlike and leaflike parts (mosses).

B. They lack true **vascular** tissue for water and food distribution and are consequently restricted to moist habitats.

C. The plant body **(thallus)** may be **dorsoventrally flattened** and **bilaterally symmetrical** (liverworts) or erect and **radially symmetrical** (mosses).

D. Like the algae, they require water for fertilization, but their reproductive organs, unlike algae, are multicellular — the **archegonium** (female) and the **antheridium** (male).

E. They have a pronounced alternation of generation cycle, in which the gametophyte is the more conspicuous and self-sustaining plant. The sporophyte is small, often ephemeral, and dependent on the gametophyte for its food and water. You may wish to review this life cycle in Figure 6.3.

III.   Class Hepaticae — Liverworts

Most liverworts live in moist shady locations; a few are essentially aquatic. *Marchantia,* the example we will study, is a flat, green, ribbonlike, dichotomously branched thallus, anchored to the substrate by rhizoids arising from the lower surface. Asexual reproduction occurs commonly by **fragmentation;** as the thallus elongates the older central portions die, leaving the tips isolated as individual new plants. Some liverworts (*Marchantia* is one) produce asexual bodies known as **gemmae.** These are pieces of thallus which have pinched themselves off and lie in a cuplike structure (**cupule** or **gemmae cup**) on the upper surface of the thallus. The gemma is a miniature thallus and, in a favorable environment (having been rinsed from the cupule by rain) it will produce a thallus exactly like the one which gave rise to it, all 1N.

Male and female plants are separate in *Marchantia*. Male plants produce **antheridial receptacles,** stalks with discs on top. Buried in these discs are multicellular globes of cells, the inner ones of which will metamorphose into *sperms* and rupture through the surface of the disc via pores when rain or dew splashes onto the mature disc. These **antheridia** are 1N already, so the sperm are formed by simple mitosis and subsequent modification for swimming.

The female plant produces an **archegonial receptacle,** but it is topped with fingerlike projections. The **archegonia,** flask-shaped structures housing the **egg,** are hung from the undersides of these projections and are shielded by fringelike **involucres.** The archegonium is also 1N and the egg is formed by modification. At first the flask is solid, then the bottom bulges out freeing the egg and cells down the center of the neck begin to degenerate, leaving an opening for the sperm to swim through. The **ventral canal cell** near the egg remains until just before the egg is ripe. If anything destroyed the egg, this cell would become a second egg. If not needed, it disintegrates.

Rainwater or heavy dew provides the water necessary for fertilization. The sperm swim along the surface film, attracted to the archegonia by chemicals secreted into the water (**chemotaxis)..** Fertilization takes place in the archegonium and the zygote remains there, nourished by its parent. A **foot** is produced at the base of the archegonium for attachment and nourishment, a short **stalk** extends from this, and the rest of the sporophyte forms a **capsule.** Inside this capsule, meiosis takes place to form numerous 1N **spores.** Among the spores, long twisted **elaters** form. In humid conditions the elaters coil tightly, but when it is dry (ideal for spore dispersal via wind) the elaters expand, pushing the spores apart and rupturing the spore case to release the spores. Half of the spores will germinate to become male plants and half to become female plants.

A.   Use a stereoscope to examine fresh and preserved specimens of *Marchantia* gametophyte thallus, gemmae cups, antheridial receptacles, and archegonial receptacles.

B.   Now, examine prepared slides of the *Marchantia* gemmae cups on a thallus, the antheridial receptacles, archegonial receptacles, and mature sporophyte.

C.   Note various types of preserved and living liverworts displayed in the lab.

IV.   Class Musci — Mosses

Like the liverworts, the typical moss has a dominant gametophyte. Both germinate from spores to form threadlike bodies called **protonemata.** These bodies soon thicken to form the stemlike and leaflike structures and rhizoids of the mature gametophyte.

Some moss plants are **monoecious** (one house), having male and female structures in the same plant. The moss we have chosen to study is, however, **dioecious** (two houses) with separate

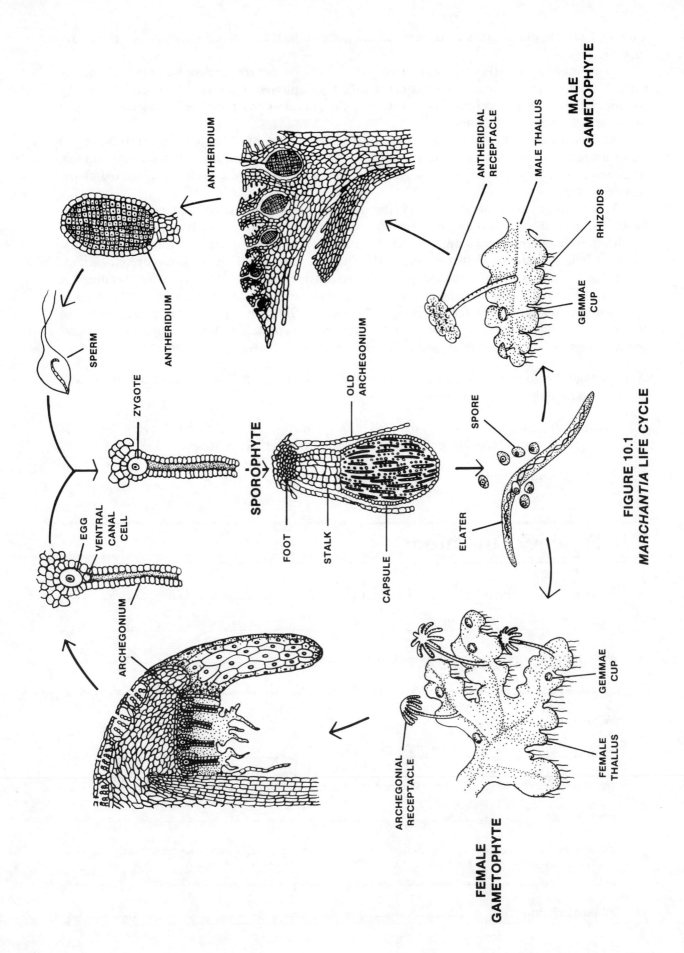

ANTHERIDIUM

ANTHERIDIAL
RECEPTACLE

MALE THALLUS

**MALE
GAMETOPHYTE**

RHIZOIDS

GEMMAE
CUP

ANTHERIDIUM

SPERM

ZYGOTE

OLD
ARCHEGONIUM

SPOROPHYTE

SPORE

ELATER

EGG

VENTRAL
CANAL
CELL

FOOT

STALK

CAPSULE

FIGURE 10.1
MARCHANTIA LIFE CYCLE

ARCHEGONIUM

ARCHEGONIAL
RECEPTACLE

GEMMAE
CUP

FEMALE
THALLUS

**FEMALE
GAMETOPHYTE**

male and female plants. It is at the top of the leafy stalk that the reproductive structures are formed.

In the male, **antheridia** very similar to those of the liverwort are formed, but stand free on the surface, surrounded by protective filaments called **paraphyses. Sperms** are produced here by mitotic cell division, just as in the liverwort, and released into rain or dew, following the chemical attractants released by the female.

At the top of the female plant, **archegonia** (shaped very much like liverwort archegonia) stand upright, surrounded by paraphyses. Here, too, there is a **ventral canal cell** for insurance. The sperm swims down the neck canal, fertilizes the **egg** and the zygote begins to grow, fed by the female parent.

This sporophyte pushes a **foot** into the parental tissue for support and nourishment and begins to elongate. At first the archegonium grows to accommodate it, but later the top is torn off by the rapidly enlarging sporophyte and its remains ride along on top of the capsule as the **calyptra.** The **capsule** is elevated by a long thin **stalk,** almost as tall as the gametophyte parent. When the capsule is mature, the calyptra falls off. Inside the capsule, meiosis has produced a number of 1N spores. Just under the caplike lid **(operculum)** are a ring of pointed flaps known as the **peristome teeth.** When humidity is high, the teeth swell and bend over the opening to retain the spores, but when conditions are dry (ideal for wind dispersal of spores) they dry out, bend back, pushing off the operculum, and allow the spores to fall away to be caught by the wind.

A. Look at the prepared slides of mass archegonia and antheridia. Find these structures as well as the paraphyses and the egg.

B. With a stereoscope, examine preserved moss gametophytes with sporophytes. Note the sporophyte's stalk, capsule, calyptra, operculum, and peristome teeth.

C. Observe various types of living and preserved mosses.

V. Review Questions

A. How can you distinguish a moss from a liverwort? _____

_____

B. Why must bryophytes live in moist environments? ~~They lack vascular~~ ~~tissue for water + food distribution~~ so must get these from environment

C. What is the function of:

gemmae Cup-like structure on upper surface of thallus; produced a thallus

paraphyses filaments that protect

protonemata form stem-like and leaf-like structures.

peristome teeth ring of pointed flaps; humidity high teeth swell when dry allow spores to fall away

elaters long twisted form; ruptures spore case to release spore to germinate

D. What is chemotaxis? Chemicals secreted in water

78

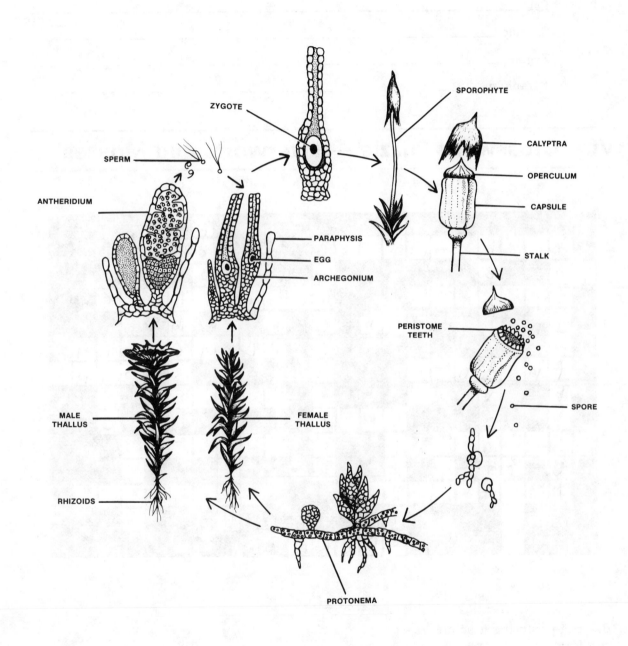

ZYGOTE

SPOROPHYTE

SPERM

CALYPTRA

ANTHERIDIUM

OPERCULUM

PARAPHYSIS

CAPSULE

EGG

ARCHEGONIUM

STALK

PERISTOME
TEETH

MALE
THALLUS

FEMALE
THALLUS

SPORE

RHIZOIDS

PROTONEMA

FIGURE 10.2
MOSS LIFE CYCLE

E.   In which generation, gametophyte (1N) or sporophyte (2N), do the following belong?

Spore _____

Calyptra _____

Archegonium _____

Protonema _____

Egg _____

Capsule _____

Zygote _____

Paraphysis _____

VI.   Crossword Puzzle — Liverworts and Mosses

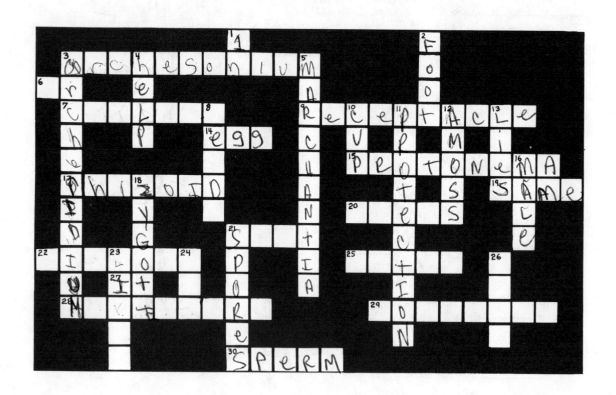

ACROSS

3.   Egg-producing structure
6.   Haploid
7.   The vegetative body of a bryophyte
9.   Structure bearing sex organs
14.   Female gamete
15.   Moss spores produce _____
17.   Rootlike structure
19.   1n and haploid have the _____ meaning.

20. Thallus liverworts possess neither _____ nor roots.
21. Moss leaves are the _____ of photosynthesis.
22. 2N
25. Necessary for bryophyte reproduction.
27. If an archegonium is a "she", and an antheridium is a "he", then a capsule is an _____.
28. A member of the Hepaticae
29. The gametophyte is the _____ generation.
30. Male gamete

DOWN

1. The nucleus of each sporophyte cell is _____.
2. Part of a bryophyte sporophyte
3. _____ receptacle
4. The sporophyte "gets a little _____ from its friends" and gametophyte.
5. A genus of liverworts
8. A moss can make neither _____ nor fruits.
10. Gemmae are often found in a _____.
11. Paraphyses surround the moss archegonia and antheridia for _____.
12. A rolling stone gathers neither a liverwort nor _____.
13. Below the operculum _____ the peristome.
16. Sperm are the _____ gametes.
18. A fertilized egg.
21. Capsule produces _____.
23. Eat a *Marchantia* and cure your _____.
24. Moss sperm swim in a film of _____.
26. Economically important moss.

KINGDOM PLANTAE
PHYLUM TRACHEOPHYTA — VASCULAR PLANTS CLASS FILICINEAE — FERNS

exercise 11

I. Objectives

After the completion of this exercise, the student should be able to do each of the following:

A. Explain what a **tracheophyte** (vascular plant) is and how it differs from all previous plants studied.

B. Identify and discuss **rhizome** and **frond** of a fern.

C. Identify fern **sporangia** with **annulus, lip cells,** and **spores.**

D. Identify fern **prothallia** bearing **archegonia, antheridia,** or a young sporophyte.

E. Discuss and draw alternation of generations in the ferns.

F. Answer the review questions at the end of this exercise.

II. Introduction to Vascular Plants

The **tracheophytes** comprise a highly diversified phylum which has exploited land habitats more successfully than any other plant group. They are characterized by a dominant sporophyte which is differentiated into **true roots, stems** and **leaves,** each made up of various types of specialized tissues. All parts of the body are interconnected by way of special conductive or **vascular** tissue which carries water and solutes throughout the plant. One type of vascular element, the evolutionary basis for all of the others, is the **tracheid,** from which the phylum takes its name. Many tracheophytes attain great size, acquiring both a highly efficient conduction system and the development of supportive mechanical tissue that enables the plant to assume an upright habit of growth.

In the more primitive tracheophytes (such as the ferns) sexual reproduction is still dependent upon water, for the sperm must swim from the male to the female sex organ. In the higher tracheophytes, (the seed plants) this swimming sperm stage is replaced by a new male gametophyte form — **pollen** — which can be dispersed by wind or by other organisms. The emergence of this type of male gametophyte was a major evolutionary breakthrough in the conquest of land, for it set seed plants free from dependence on water for completion of their life cycles.

III.   Class Filicineae — Ferns

The fern structure with which we are most acquainted is the vascular sporophyte. Ferns, once the dominant vegetation during the "coal-making age" approximately 300 million years ago, are distributed throughout the world in various forms and shapes. Some ancient ferns attained great size, but today only a relatively few tropical forms remain treelike. The sporophyte can exist indefinitely, spreading by underground stem branches, **rhizomes,** without passing through the sexual stage.

A.   Examine the various herbarium samples of fern types, noting the variation in the **frond** (leaf) of several species. There is a general trend toward greatly dissected compound leaves. Note the reproductive structures of species where present. In some ferns the sterile (photosynthetic) frond and the fertile fronds are completely separate, while in others, the sterile and fertile leaves are either on the same leaf stalk, or the reproductive structures are borne on the undersides of the photosynthetic tissues.

Distinguish between the rhizome, the **leaf stalk, blade** (broad green part) and a **pinna** (leaflet) of a fern frond, and examine the pattern of **sori** (sporangium clusters) distribution of several ferns.

B.   Examine slides of fern sori and sporangia. Within these **sporangia** meiosis occurs to produce 1N spores. The row of cells down the back of the helmet-like sporangium is called the **annulus.** Since these cells are thin-walled on the outer edge, water loss during spore maturation causes this row to shrink, eventually causing the sporangium to crack along the large **lip cells** and catapult the spores away from the parent plant.

C.   Upon germination those spores do not grow directly into the leafy fern seen on the herbarium sheets. The first strand of cells to emerge is thread-like and is known as the **protonema.** Later, there are divisions in two planes, producing a flat heart-shaped or bi-lobed structure called the **prothallium** — the mature gametophyte plant. **Rhizoids** arise on the under surface, as do the reproductive organs.

The **antheridia** are produced first, release **sperm** and then drop off. Then the **archegonia** are produced. This staggered production of gametes promotes cross fertilization between the prothallia of different monoecious parent plants. Some fern species are dioecious.

Obtain a slide of fern **prothallia** which show gametophytes in both sperm and egg producing stages.

D.   The young **zygote** develops in the archegonium and for a short time takes nourishment from the gametophyte. Soon the **sporophyte** takes on the recognizable leafy-fern look, crushing the prothallium as it grows.

E.   OPTIONAL — Observe various other groups of plants (not members of the class Filicineae) whose life cycles are similar. These "fern allies" include *Equisetum* (Horsetail), *Selaginella,* and *Lycopodium* (Club Moss).

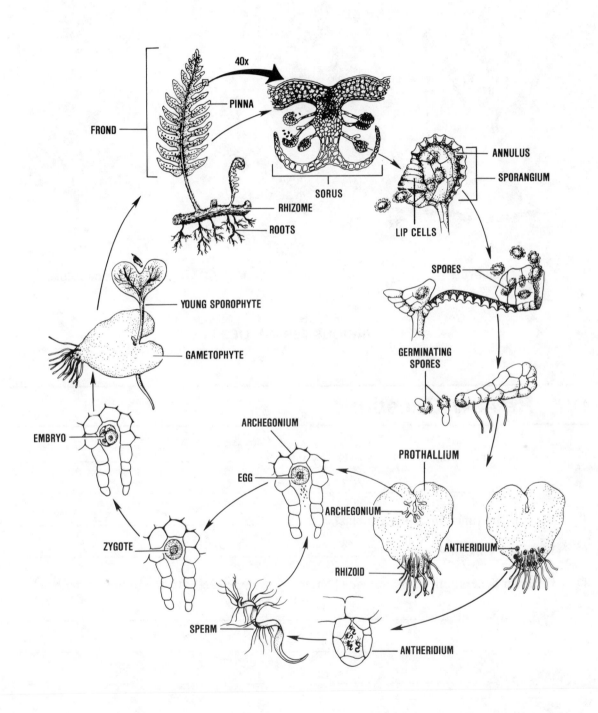

FROND

PINNA

40x

SORUS

RHIZOME

ROOTS

ANNULUS

SPORANGIUM

LIP CELLS

SPORES

GERMINATING
SPORES

YOUNG SPOROPHYTE

GAMETOPHYTE

PROTHALLIUM

EMBRYO

ARCHEGONIUM

EGG

ARCHEGONIUM

ANTHERIDIUM

RHIZOID

ZYGOTE

SPERM

ANTHERIDIUM

**FIGURE 11.1
FERN LIFE CYCLE**

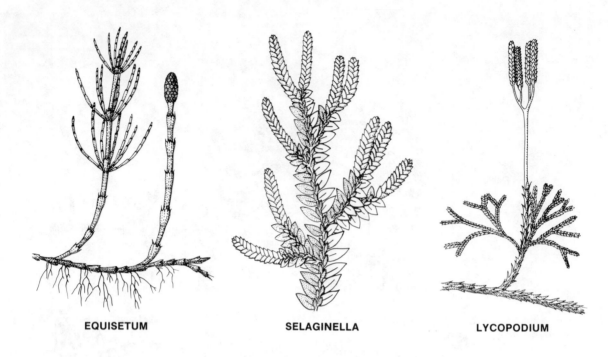

EQUISETUM   SELAGINELLA   LYCOPODIUM

**FIGURE 11.2
VARIOUS FERN ALLIES**

IV. Review Questions

A. What are the advances of ferns over bryophytes? _____

_____

B. What is the function of the annulus? _____

C. What is a rhizome? _____

D. In which generation, gametophyte (1N) or sporophyte (2N), do the following belong?

prothallium _____

pinna _____

spores _____

frond _____

sporangium _____

antheridium _____

86

KINGDOM PLANTAE
PHYLUM TRACHEOPHYTA — VASCULAR PLANTS
CLASS GYMNOSPERMAE

exercise 12

I. Objectives

After the completion of this exercise, the student should be able to do each of the following:

A. Describe the life cycle of the pine as an example of a **gymnosperm.**

B. Identify various cones and determine whether they produce **pollen** or **ovules.**

C. Discuss the adaptations the pine has made to cold and dry environments.

D. Answer the review questions at the end of this exercise.

II. Gymnosperms

Although much of the earth's surface is covered by conifers, the number of species is relatively few. Many groups of the class are complete or partially extinct. We will study *Pinus,* a pine, which is one of the more modern forms of gymnosperm. Pines and most other living gymnosperms are **conifers.** They bear cones.

The conifers include our well-known evergreens: pine, spruce, fir, hemlock, redwood, as well as the southern conifers. They include the largest and oldest of all living things known. A bristlecone pine growing on the rim of the Grand Canyon has been dated as being over 2000 years old and is the oldest known living thing on earth.

In gymnosperms, the gametophytes live as parasites on sporophyte plants. Sporophytes which bear male gametophytes on one plant and female gametophytes on another are termed **dioecious** (= two houses). Many gymnosperms are dioecious, but more are **monoecious** (= one house), bearing male and female gametophytes on the same sporophyte plant.

The reproductive structures (sporangia) are located on specialized leaves called **sporophylls.** In both gymnosperms and angiosperms two types of spores are produced. Four large **megaspores** (1N) are meiotically produced in a **megasporangium** or **ovule.** Three of the four megaspores die, leaving one megaspore to grow and become the female gametophyte. This gametophyte contains two or three archegonia, each containing one egg.

Small, numerous **microspores** (1N), the second type of spore, are meiotically produced in a **microsporangium.** Then, these microspores or **pollen** undergo mitosis to form two nuclei; one, the **generative nucleus,** will form two sperm; the other, the **tube nucleus,** will synthesize the **pollen tube.**

The **megasporophylls** are grouped together in clusters called **cones,** as are the **micro-sporophylls.** The **ovulate cones** (megasporangiate) are borne on the sides of the branches and last one to two years before opening and dropping. This is the kind of cone usually meant when one says "pine cone." The **pollinate cones** are produced at the tips of the branches and last only a few months.

The wind blows the pollen into the cracks of the female cone and the pollen tube grows toward the interior. The tube nucleus works itself to death making cytoplasm, while the generative nucleus follows it down the tube, eventually dividing into two **sperm** nuclei. Just outside the archegonia the pollen tube ruptures and the sperm swim through the cytoplasm to the eggs. Although all eggs may be fertilized, usually one embryo will grow faster than the others and the weaker embryos will abort — survival of the fittest. The surviving embryo produces 6, 8 or more **cotyledons** to store food, then goes dormant. The gametophyte tissue is dissolved and absorbed into these cotyledons. The surrounding megasporophyll tissue dries and flattens to produce a winged **seed coat.** When the seed is blown from the opened female cone and germinates, it will use the food stored in the cotyledons to supply nourishment until it can produce roots and leaves of its own.

A.   Learn the life cycle of the pine. (See Figure 12.1)

B.   Examine prepared slides of ovulate and pollinate cones of *Pinus.*

C.   Prepare a slide of pine pollen. Notice the air bladders on both sides of the pollen grain.

D.   Observe various preserved materials which include immature and mature ovulate cones, pollinate cones, and pine seeds.

E.   On demonstration is a slide showing pine ovules. Be able to identify the female gametophyte, the archegonia, and the megasporangial wall.

F.   On demonstration is a slide showing a pine seed. Identify the embryo, the gametophyte, and the seed coat.

G.   The pine needle or leaf is a result of adaptation to cold and dry climates. During winters, frozen water is useless to plants. Therefore, antifreezing resin is made and stored in **resin canals** of the leaf, stem, and root of the pine. In addition, resin has insecticidal and fungicidal properties.

   1.   Obtain a prepared slide of a cross-section of a pine needle. Observe the **hypodermis,** a thick layer of supporting and water-retaining cells beneath the **epidermis,** the numerous, sunken pairs of **guard cells** which regulate gas exchange and also prevent excessive water loss, the photosynthetic **mesophyll,** the waterproofing **endodermis** surrounding the vascular tissue, the thick-walled **xylem,** which conducts water and minerals, and the **phloem,** which carries food and hormones.

   2.   Label the following illustration of a cross-section of the pine needle with the terms given above.

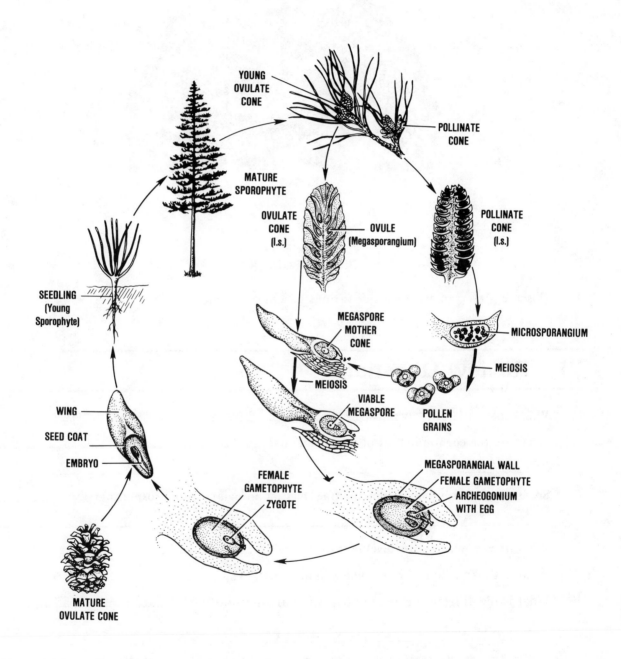

YOUNG
OVULATE
CONE

POLLINATE
CONE

MATURE
SPOROPHYTE

OVULATE
CONE
(l.s.)

OVULE
(Megasporangium)

POLLINATE
CONE
(l.s.)

SEEDLING
(Young
Sporophyte)

MEGASPORE
MOTHER
CONE

MICROSPORANGIUM

MEIOSIS

MEIOSIS

VIABLE
MEGASPORE

POLLEN
GRAINS

WING

SEED COAT

EMBRYO

FEMALE
GAMETOPHYTE

ZYGOTE

MEGASPORANGIAL WALL
FEMALE GAMETOPHYTE
ARCHEOGONIUM
WITH EGG

MATURE
OVULATE CONE

FIGURE 12.1
PINE LIFE CYCLE

89

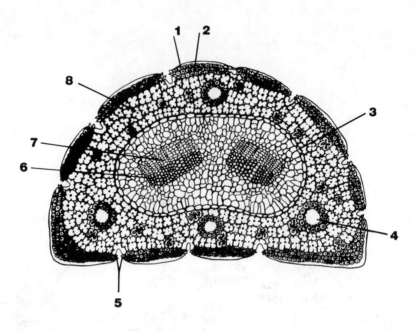

FIGURE 12.2
PINE NEEDLE (X.S.)

H.   Observe preserved specimens of various genera of gymnosperms.

III.   Review Questions

A.   What does the word "gymnosperm" mean? _____

B.   What are the economic uses of gymnosperms? _____
_____

C.   Seeds are found in which type of cone, megasporangiate or microsporangiate? _____
_____

D.   What does a microspore develop into? _____

E.   What is the function of resin canals in pine needles? _____

F.   What is the difference between pollination and fertilization? _____
_____
_____

G.   What is another name for a megasporangium? _____

KINGDOM PLANTAE
PHYLUM TRACHEOPHYTA — VASCULAR PLANTS
CLASS ANGIOSPERMAE
SUBCLASS DICOTYLEDONAE;
SUBCLASS MONOCOTYLEDONAE

exercise 13

I. Objectives

After the completion of this exercise, the student should be able to do each of the following:

A. Identify the parts and their functions of a typical **angiosperm flower, fruit,** and **seed.**

B. Describe the life cycle of an angiosperm.

C. Identify an angiosperm as a **dicot** or a **monocot** on the basis of number of flower parts, number of cotyledons, and pattern of leaf venation.

D. Examine a flower or fruit and determine the **carpel number.**

E. Explain the **carpel theory.**

F. Answer the review questions at the end of this exercise.

II. Floral Anatomy

Although the flowers of angiosperms show a tremendous diversity (80% of all species of plants) the structures which make up the flower are basically the same throughout. The following parts are recognized although not all may occur in any one flower:

The supporting stalk of the flower is called the **pedicel.**

A **receptacle** is at the center base of a flower and represents the end of the stem. In some cases it may form a dome for the other appendages or a cup with some appendages on the rim and others on the bottom. It is not usually a large or very noticeable part of most flowers.

The **calyx,** composed of **sepals,** makes up the lower or outermost whorl of floral appendages. Usually these are small and leaf-like, although different in shape from the other foliage leaves of the plant. Their major function is the protection of the immature inner parts, but in some cases they look and function as petals as well (lily).

The **corolla,** composed of **petals,** arises inside and usually above the sepals, usually being brightly pigmented and very broad. These serve as additional protection in early stages and later to attract pollinators. Some serve as "landing pads" for the pollinator, and may have glands (nectaries) at the base which produce sweet nectar and/or volatile oils to provide the pollinator with a scent trail to the flower. (Those angiosperms which are wind pollinated usually do not have petals.)

The male reproductive structures are the **stamens,** arising inside the petals. Each consist of a slender stalk or **filament** and variously shaped, enlarged sacs called **anthers.** Inside the anther, special groups of cells undergo meiosis to produce male gametophytes — **pollen grains.**

One or more **pistils** may occur in the center of the flower. A pistil is composed of: (1) a **stigma** which produces a sticky substance which catches pollen grains and induces their germination; (2) a **style** through which the pollen tubes grow toward the ovules; and, (3) an **ovary** — the basal region which contains one or more ovules (megasporangia). It is within the ovule that one of the four megaspores will produce a female gametophyte. In angiosperms the female gametophyte contains eight nuclei at maturity and is called an **embryo sac.**

A.   Label the flower illustration with these terms: anther, filament, ovary, ovule, pedicel, petal, receptacle, sepal, stamen, stigma, and style.

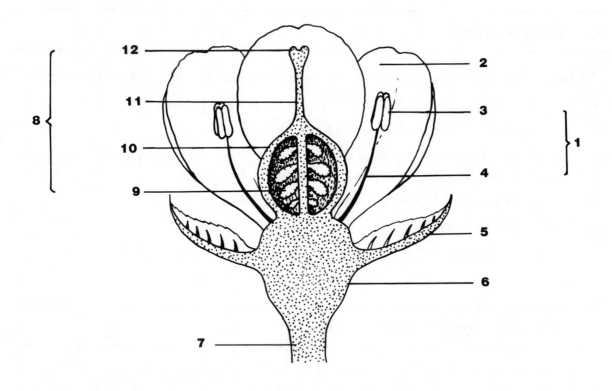

FIGURE 13.1
GENERALIZED FLORAL ANATOMY

B.   At fertilization the egg fuses with one of the two sperm produced by a pollen grain. A zygote (2N) results. However, the remaining sperm is not wasted but unites with the two polar nuclei of the embryo sac to form a triploid (3N) organism called the **endosperm.** The endosperm will devour the remaining embryo sac and provide food for the young (2N) **embryo.** Learn the following diagram of the typical angiosperm life cycle.

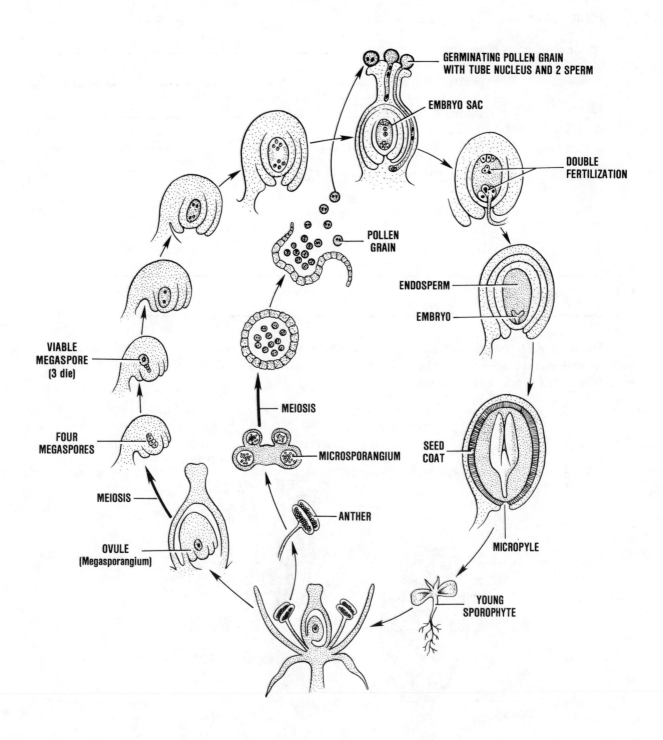

FIGURE 13.2
ANGIOSPERM LIFE CYCLE

Labels in figure:
- GERMINATING POLLEN GRAIN WITH TUBE NUCLEUS AND 2 SPERM
- EMBRYO SAC
- DOUBLE FERTILIZATION
- POLLEN GRAIN
- ENDOSPERM
- EMBRYO
- VIABLE MEGASPORE (3 die)
- MEIOSIS
- SEED COAT
- FOUR MEGASPORES
- MICROSPORANGIUM
- MEIOSIS
- ANTHER
- MICROPYLE
- OVULE (Megasporangium)
- YOUNG SPOROPHYTE

C.   Dissect a fresh or preserved flower using dissecting utensils and a stereomicroscope. Fill in the following information concerning the dissected plant flower:

Plant name _____

Number of parts: sepals _____

petals _____

stamens _____

pistils _____

D.   Examine prepared slides of the lily ovary and flower buds.

III.   Fruit

The **fruit,** a unique characteristic of the angiosperms, develops from the ovary of the flower and sometimes includes the receptacle. Study the various fruits on demonstration and carefully note from what types of flowers they arose.

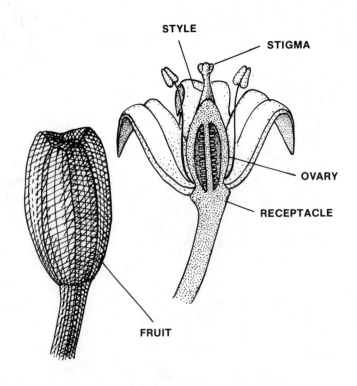

FIGURE 13.3
LILY FLOWER AND FRUIT

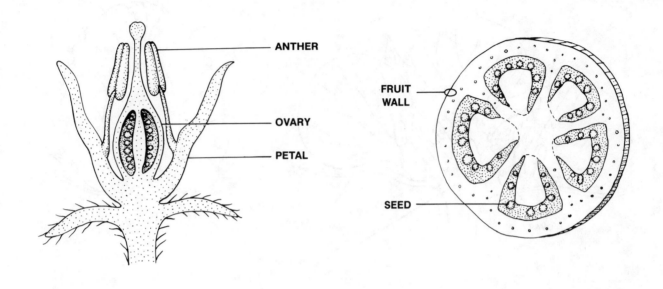

FIGURE 13.4
TOMATO FLOWER AND FRUIT

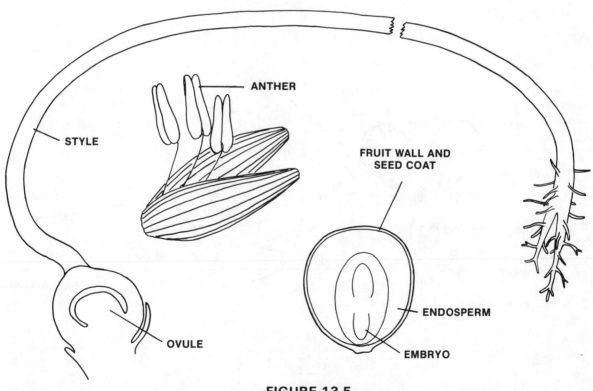

FIGURE 13.5
CORN FLOWERS AND FRUIT

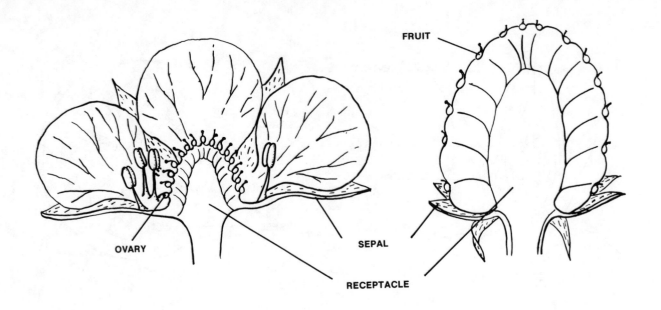

FIGURE 13.6
STRAWBERRY FLOWER AND FRUIT

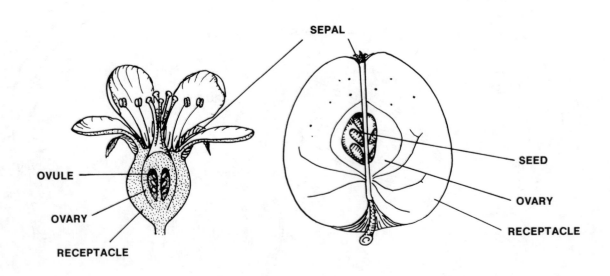

FIGURE 13.7
APPLE FLOWER AND FRUIT

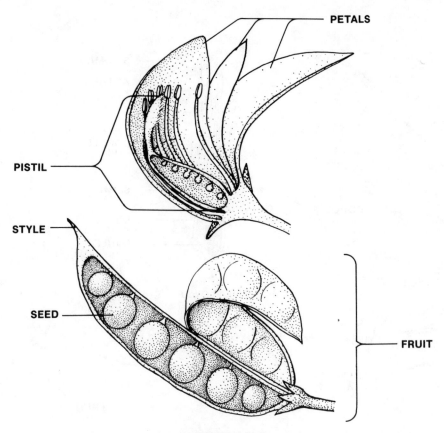

PETALS

PISTIL

STYLE

SEED

FRUIT

FIGURE 13.8
PEA FLOWER AND FRUIT

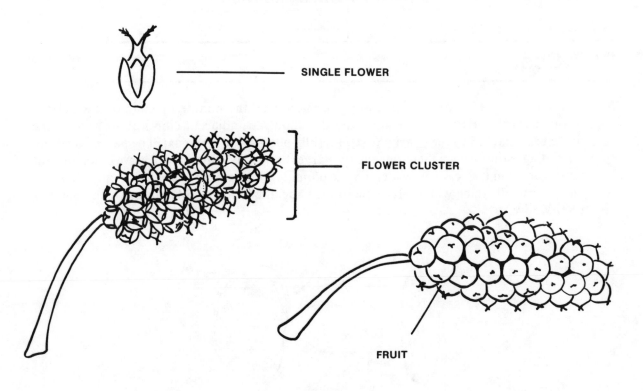

SINGLE FLOWER

FLOWER CLUSTER

FRUIT

FIGURE 13.9
MULBERRY FLOWERS AND FRUITS

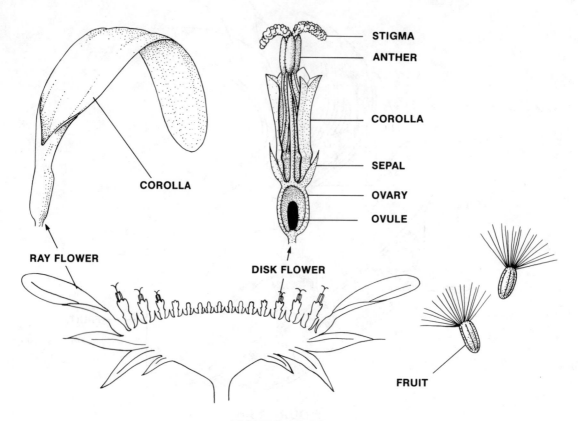

FIGURE 13.10
COMPOSITE FLOWER AND FRUITS

IV.   Seed

A.   Examine a soaked bean seed. Identify the **micropyle,** the minute opening on the surface through which the pollen tube grew and the **hilum,** the adjacent elliptical area where the ovule was attached to the ovary. Next, carefully peel off the **seed coat.** The part of the bean remaining is the embryo. Separate the two large fleshy **cotyledons** which serve as an area for food storage in the bean. Now identify the **plumule** which will produce new stem and leaves and the **radicle** which will produce the roots. Label the following illustration with the terms given above.

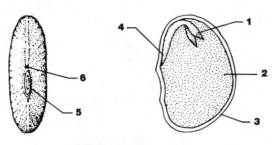

FIGURE 13.11
BEAN SEED

B. Observe bean seedlings on demonstration. Identify which parts of the seedling came from the embryonic structures mentioned previously.

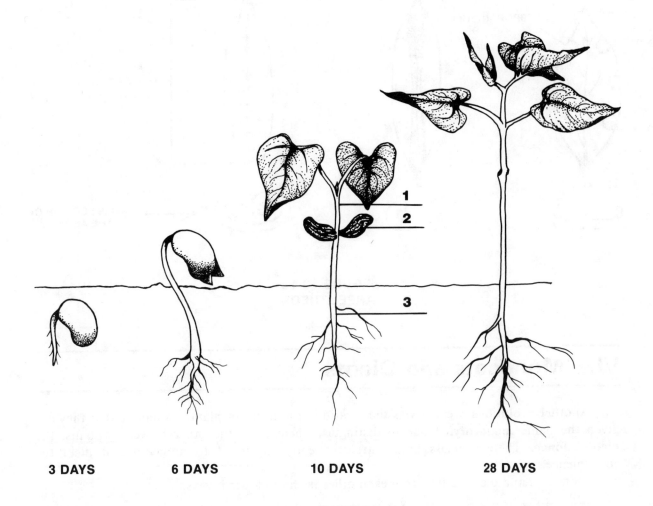

3 DAYS   6 DAYS   10 DAYS   28 DAYS

FIGURE 13.12
BEAN GERMINATION

V. Carpel Theory

The evolution of angiosperms is still a mystery to scientists. However, it is believed that they descended from plants whose sporangia were borne on the edges of leaves. Then, due to the increasing selective pressures by the rapidly-evolving insects, the leaf slowly wrapped around and enclosed the megasporangia. This leaf today is known as a **carpel**, and the ovules are no longer naked but are covered (angiosperm). (See Figure 13.13.)

The numerous carpels in the flower through millions of years of evolution often fused in various groupings to form the pistil. Thus, some flowers have pistils containing one carpel while other pistils of flowers have numerous carpels. Reexamine several fruits and flowers by viewing their cross sections or counting the stigmata. How many carpels do they have?

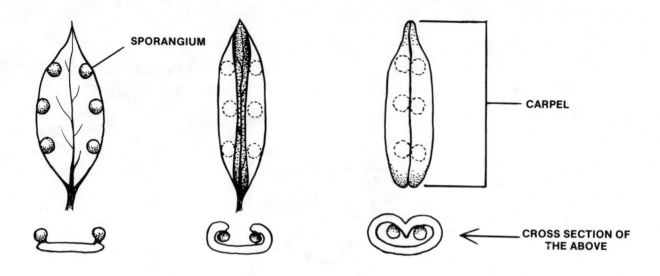

FIGURE 13.13
CARPEL THEORY

VI. Monocots and Dicots

Another evolutionary event was the creation of a group of plants so unique that they are given the name **monocotyledonae** to distinguish them from the other flowering plants, the **dicotyledonae.** These subclass names are frequently shortened to **monocot** and **dicot** for convenience.

Monocots and dicots differ from each other in the following ways:

	MONOCOTS	DICOTS
number of cotyledons	one	two
venation of leaves	parallel	netted
root system	usually fibrous	either fibrous or taproots
flower parts	3's or multiples of 3	4 or 5, or multiples of 4 or 5
stems	mostly herbaceous	herbaceous or woody

In the next exercise, you will also see differences in the internal anatomy of the roots and stems of monocots and dicots.

VII. Review Questions

A. Angiosperms compose what percentage of all the plant kingdom? _____

B. What structure of a flower is an immature fruit? _____

C. The embryo sac (mature female gametophyte) typically has _____ nuclei.

100

D. **Double fertilization** in the angiosperms results in the zygote and the ————————— .

E. A lily (monocot) has its flower parts in multiples of ————————————— .

F. What structure does the radicle in a seed produce in the mature plant? —————————
——

G. What does "angiosperm" mean? ————————————————————————————

H. One or more carpels make up: a. a pistil   b. a corolla   c. a calyx   d. a cone

I. What does the term "dicot" actually mean? ———————————————————————

VEGETATIVE STRUCTURES OF ANGIOSPERMS

I. Objectives

After the completion of this exercise, the student should be able to do each of the following:

A. Describe the external characteristics of the root, stem, and leaf.

B. List the functions of **root, stem,** and **leaf.**

C. Explain how the **root, apical,** and **cambium meristems** function.

D. Identify the internal structures of roots, stems, and leaves.

E. Explain what causes **growth rings** and determine the age of a tree from a section of its trunk.

F. Name the various types of roots based on their origin or structure.

G. Explain the differences between a monocot and dicot stem.

H. Answer the review questions at the end of this exercise.

II. Roots

The major functions of roots are: absorption of water and minerals, anchorage, and storage of food in the form of starch and other carbohydrates. Roots have few distinctive external features except root hairs for absorption and a root cap. The **root cap** protects the meristem as the root pushes through the soil. The root cap cells are replenished internally as they are sloughed away on the outside.

The **root meristem,** like all meristematic regions, has the property of **totipotency,** the capacity for continuous division and growth. It is in this area where new root and root cap cells are produced by mitotic cell division. As these immature cells begin to grow they elongate. Eventually these cells differentiate into their predetermined form and functions. Below, the meristem remains intact and busily produces new cells.

A. Get a prepared slide of an onion *(Allium)* root tip. Note the root cap, root meristem, and region of elongation.

B. On demonstration are roots which possess **root hairs,** minute structures for absorbing water and soil minerals.

C. Label the illustration below with these terms: region of elongation, region of maturation, root cap, root hair, and root meristem.

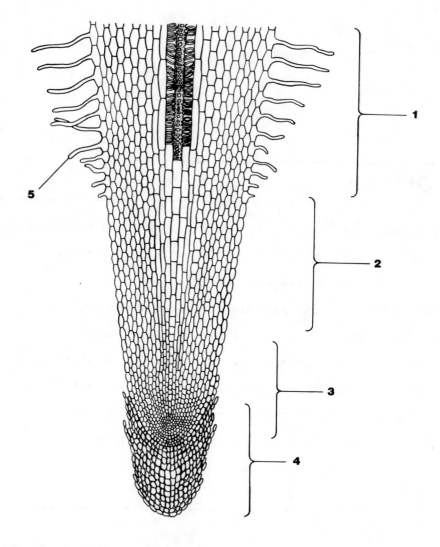

FIGURE 14.1
ROOT TIP (L.S.)

D. In the area of maturation, the root cells are in their final form. The root is covered by a protective cell layer, the **epidermis.** It is this epidermal layer which also produces outgrowths of its cell walls and cytoplasm called root hairs.

Inside the epidermis, the **cortex** of a dicot root shows its storage function in the many pinkish starch grains present in many of the cells. This tissue also helps transport water and minerals from the root hairs to the xylem. The innermost layer of the cortex is the **endodermis,** a single layer of thick-walled cells which act as waterproofing and support tissue for the vascular cylinder. Just inside the endodermis is a **pericycle** of one or more cells in thickness. This layer is important because it produces the secondary or branch roots and the vascular cambium of the root at a later time.

Surrounded by the pericycle is the vascular tissue or **stele,** composed of xylem and phloem. The xylem cells, when mature, die and become hollowed cell walls (similar to drinking straws) in order to conduct water with dissolved minerals upward to the leaves. In the typical dicot root, the xylem is in the center and around the xylem are bundles of **phloem.** Phloem cells conduct food and hormones. Phloem consists of large **sieve tube cells** through which solutes are conducted and the **companion cells** which maintain and control the sieve tube cells. Since the food and hormones are conducted via active transport, phloem cells must remain alive in order to function. Between the xylem and phloem a few inconspicuous meristematic cells occur. This is the **cambium,** which may eventually manufacture new secondary xylem and phloem.

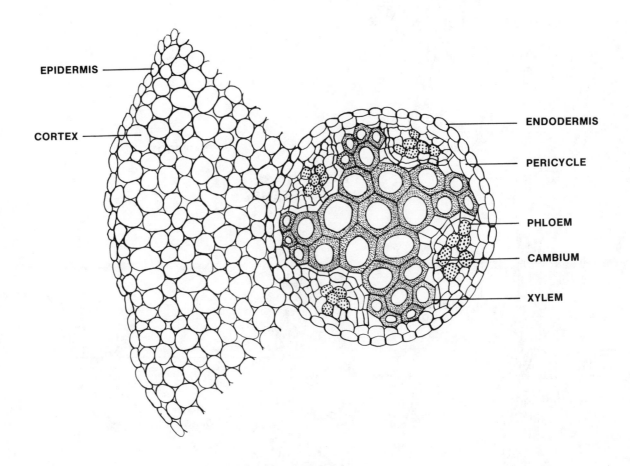

FIGURE 14.2
MATURE DICOT ROOT (X.S.)

E. Look at the demonstration slide showing lateral or branch root origin. Remember, the pericycle becomes temporarily meristematic and produces these secondary branch roots.

F. Roots can be classified structurally either as **fibrous** or **taproots.** They can also be grouped according to their origin; **primary roots** arise from the radicle of the embryo, **secondary roots** arise from the pericycle, and **adventitious roots** arise from the stem or leaves. Study the various types of roots on demonstration.

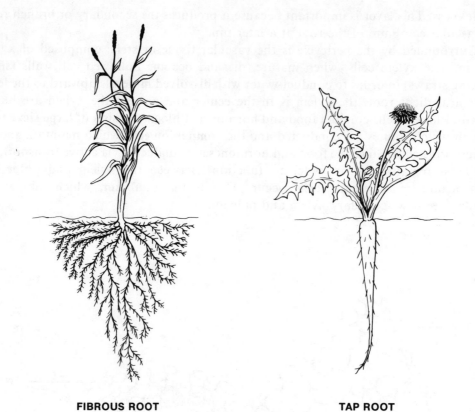

FIBROUS ROOT

TAP ROOT

FIGURE 14.3
ROOT SYSTEMS

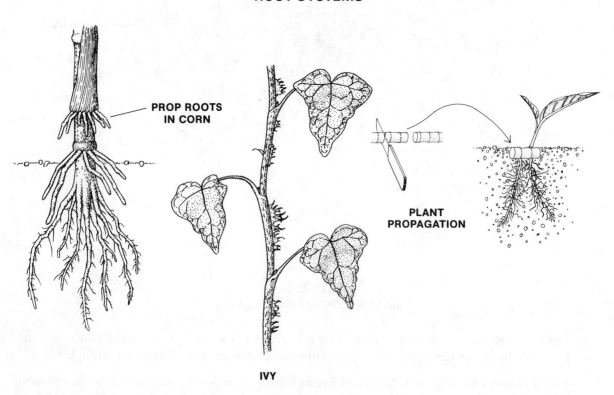

PROP ROOTS
IN CORN

PLANT
PROPAGATION

IVY

FIGURE 14.4
ADVENTITIOUS ROOTS

III. Stems

The major stem functions are: conduction of food, water, and minerals between root and stem; production and display of photosynthetic leaves; and storage of food.

A. Stems have several distinct external features which you should examine and be able to identify using the following illustration as a guide. Leaves and buds arise at the **nodes.** The areas in between nodes are called **internodes.** Buds are immature stems, those which arise along the sides of the stem producing branches of the stem, and those located at the tips of stems producing stem elongation. These are known as **lateral** and **terminal buds,** respectively. **Terminal bud scars** are formed each spring when the scales covering the **terminal bud** fall off. When leaves drop off in the fall, they leave **leaf scars** as well. Stems are also covered by a number of light or dark spots. These disruptions in the bark are called **lenticels,** and function in gas exchange.

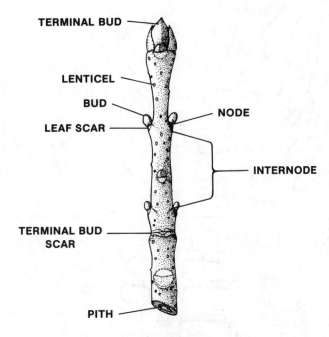

FIGURE 14.5
FEATURES OF A DORMANT TWIG

B. The shoot apex is covered by young leaves which arch over the meristem. This is more difficult to see in fresh material, so we will use a longitudinal section of an *Elodea* bud. Obtain such a slide and try to identify the parts labeled in the following illustration.

There is no root cap and the zones of elongation and differentiation are not as conspicuous as in the root. At about ground level the core of vascular tissue of the root branches out to form numerous vascular bundles. This leaves a large parenchyma core called the **pith** in the center. *Elodea* is a rooted plant even though it is most often seen as stem cuttings. Being an aquatic plant it has numerous air chambers in the pith and the cortex to aid in flotation.

At the extreme tip is the primary meristem of the stem. On the flanks of this tall meristem are the leaf **primordia** appearing as bumps in young stages. As the leaves mature another "bump" appears where the leaf joins the stem. These are the **axillary buds** and may become new vegetative or flowering stems.

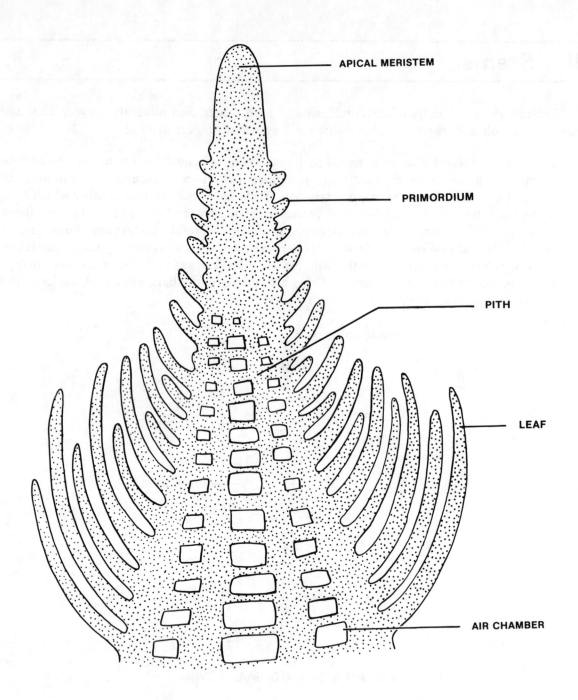

APICAL MERISTEM

PRIMORDIUM

PITH

LEAF

AIR CHAMBER

FIGURE 14.6
ELODEA **BUD (L.S.)**

C. Study a prepared slide of a cross-section of both a monocot and a dicot stem.
    The monocot stem has a scattered arrangement of the vascular bundles. The individual vascular bundles resemble faces, with xylem for "eyes, nose, and mouth," phloem for the "forehead," and supporting cells for "hair and beard." Surrounding the vascular bundles is pith, an area of large, white parenchyma cells. Although parenchyma cells are found in many areas of a plant (root cortex, epidermis, and leaf mesophyll) and have various functions, the function of the pith parenchyma cells is storage of food. Sugar cane contains sucrose in its pithy stem. Enclosing the monocot stem is the protective epidermis.

108

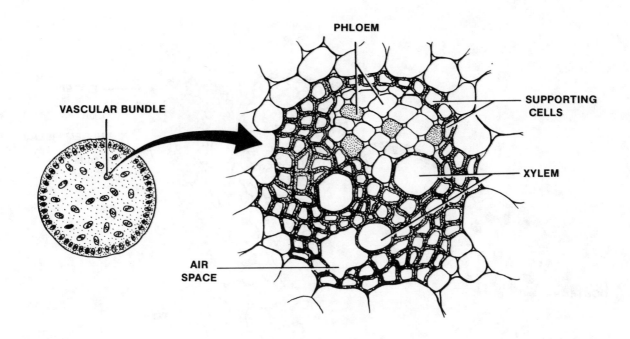

VASCULAR BUNDLE

PHLOEM

SUPPORTING CELLS

XYLEM

AIR SPACE

FIGURE 14.7
MONOCOT STEM (X.S.)

D. In the young dicot stem (usually the smaller one on the slide) the outermost layer of cells is the **epidermis.** These cells have a waxy, non-living covering called the **cuticle.** Just inside is a narrow, parenchymous **cortex** for storage. The stele of the stem is broken up into vascular bundles with **pith rays** running between them and the large, central **pith.**

   The inner portion of the vascular bundles is the **xylem,** separated from the phloem by a small area of the meristematic **cambium.** Just outside, the conducting phloem area is a cap of **phloem fibers** for support of the young stem.

E. In many of the dicots the plant lives more than one or two years and the stem usually becomes woody. The cambium stretches to connect the vascular bundles and begins to manufacture new xylem and phloem. In the areas of the world with seasonal climates, the growth of the vascular cambium changes likewise. In the spring, warm temperatures and plentiful rainfall and sunshine produce a need for numerous water-conducting **vessels** (an advanced type of xylem cell shaped similarily to a plumbing pipe). As summer begins, the weather, although hot, brings fewer inches of rain. Thus, the cambium manufactures fewer and fewer vessels. Then in the autumn, the leaves stop photosynthesis and fall from the trees. No wood is produced in the fall or winter.

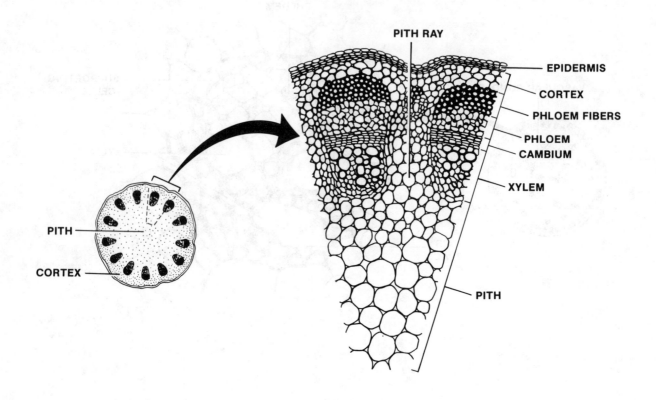

**FIGURE 14.8
YOUNG DICOT STEM (X.S.)**

The contrast between the larger and lighter-shaded **"spring wood"** and the smaller, darker **"summer wood"** results in **tree rings.** In wet years, the growth rings will be wide; in drier years, these rings will be narrow. Thus, examination of tree rings can trace the condition of past years in the location where the tree grew. Radioactive carbon dating is often checked and calibrated with the results of tree ring analysis.

Pick out a *Tilia* 2-year stem slide and find all of the parts labeled on the following diagram.

110

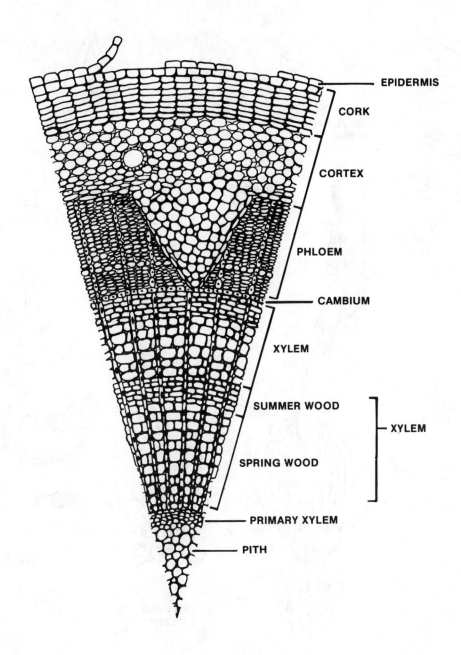

EPIDERMIS

CORK

CORTEX

PHLOEM

CAMBIUM

XYLEM

SUMMER WOOD

XYLEM

SPRING WOOD

PRIMARY XYLEM

PITH

FIGURE 14.9
TILIA **TWO-YEAR STEM (X.S.)**

IV.   Leaves

A.   The major function of leaves is usually photosynthesis. However, other minor functions may be protection, secretion, storage of food, or even the capture of small animals for nitrogen. Note the various leaves on display.

111

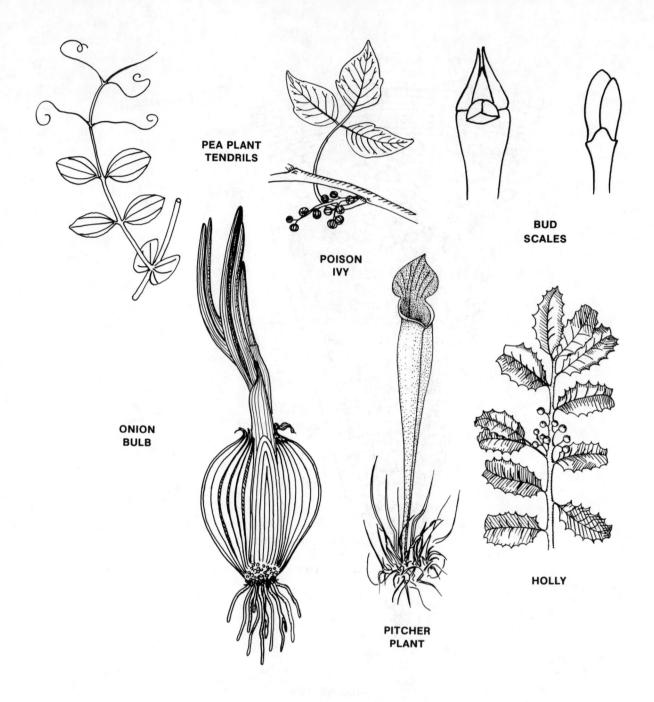

PEA PLANT
TENDRILS

POISON
IVY

BUD
SCALES

ONION
BULB

PITCHER
PLANT

HOLLY

FIGURE 14.10
LEAVES SPECIALIZED FOR FUNCTIONS OTHER THAN PHOTOSYNTHESIS

B. Obtain a prepared slide of a cross-section of a typical dicot leaf. Locate the single-layered **epidermis,** noting those cells of the upper epidermis have a waterproofing **cutin** covering. Now locate the **guard cells** and the **stomates** on the lower epidermis.

Between the epidermis, the mesophyll parenchyma is found. There are one or more layers of closely packed **palisade parenchyma** just below the upper epidermis, where most of the photosynthesis takes place. Below this is an area of loose **spongy parenchyma,** providing numerous air chambers for gas exchange via the stomata.

In the center of the leaf is the major **vein,** or vascular bundle. Such veins branch out through the parenchyma to distribute needed water and minerals and pick up sugars.

Farther out from the midvein you may see streak-like clusters of cells. This is where the slice caught a vein curving out to the edge of the leaf. You may be able to see the rings of the xylem vessels very well in one of these side veins. Each such vein contains the **xylem** for water and mineral transport and the **phloem** for food and hormone distribution and transport. Xylem will be the large redwalled cells on the top of the vein, while the phloem appears as densely cytoplasmic, greenish, and on the bottom of the vein.

C. Label the illustration of the dicot leaf cross-section with these terms: cuticle, epidermis, palisade parenchyma, spongy parenchyma, stomate, vascular bundle, guard cells, and air space.

D. Detach a *Zebrina* ("Wandering Jew") leaf from its stem and note the purple pigmentation on the lower epidermis. Place a small section of this leaf, purple side up, on a slide. Examine under a microscope to notice the numerous stomates which appear as cat's eyes glowing in a field of purple. Draw what you see.

If *Zebrina* leaves are not available, examine a prepared slide of the lower epidermis of a *Transcendia* leaf and locate the stomates and their surrounding guard cells.

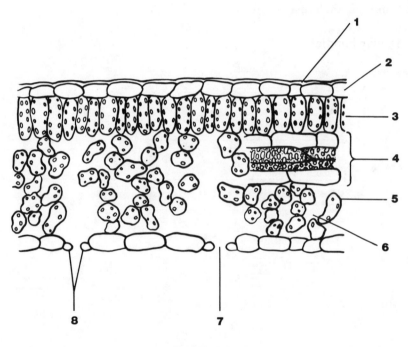

FIGURE 14.11
DICOT LEAF (X.S.)

VI. Review Questions

A. What is totipotency? _____

_____

Name at least 4 areas in plants which possess totipotency. _____

_____

B. Epidermal outgrowths for absorption in roots are _____

C. Roots arising from an African violet leaf can be termed _____

D. What is the function of lenticels? _____
_____

E. If monocots possess no cambium, can they produce wood? _____

F. In southern California where the winters are warm and rainy, while the summers are hot and dry, would trees produce growth rings? _____

    Why? _____

G. In tropical rainforests with constant rainfall throughout the year, do the trees produce prominent growth rings? _____ Why? _____
_____

H. The major areas of photosynthesis in an angiosperm leaf are:
    a. epidermis
    b. palisade parenchyma
    c. spongy parenchyma
    d. only b & c
    e. vascular bundles

KINGDOM PROTISTA

I. Objectives

After the completion of this exercise, the student should be able to do each of the following:

A. Give the distinguishing characteristics of the Kingdom Protista.

B. Explain how protozoans differ from other protists, and list the phyla included as protozoans.

C. Identify the following organisms from prepared slides and give the phylum to which each belongs:

> *Trypanosoma gambiense*   *Paramecium caudatum*
> *Amoeba proteus*   *Plasmodium malariae*
> foraminifera   radiolaria
> diatoms   *Euglena* sp.
> *dinoflagellates*

D. List the distinguishing characteristics of the following protist phyla:

> Phylum Chrysophyta   Phylum Zoomastigina
> Phylum Euglenophyta   Phylum Ciliophora
> Phylum Sarcodina   Phylum Sporozoa
> Phylum Pyrrophyta

E. Identify and give the functions of each of the following structures:

flagella	pellicle	contractile vacuole
cilia	eyespot	paramylum granule
pseudopodia	oral groove	undulating membrane
micronucleus	macronucleus	food vacuole
oral groove	gullet	anal pore
raphe	trichocysts	

F. Identify in prepared slides and explain the processes of **conjugation** and **transverse binary fission** in *Paramecium*.

115

G. Compare feeding and nutrient processing in *Paramecium* and *Amoeba*.

H. Answer the review questions at the end of this exercise.

II. Introduction

The Protist Kingdom is indeed a very diverse group of organisms. The only unifying characteristics in this group are that all its members are both eukaryotic and unicellular. The first three phyla which we will consider in this exercise are photosynthetic and have other plant-like traits. The others are all heterotrophic and are often grouped together as **protozoans** (= first animals) because of their animal-like characteristics.

III. Phylum Pyrrophyta

The members of the Phylum Pyrrophyta (= fire plants) are characterized by their possession of a cell wall composed of **cellulose plates** (when they have a cell wall) and by their possession of two flagella — one directed posteriorly, and the other, laterally. Many members of this group luminesce when disturbed, causing the water to "sparkle" at night, hence the name "fire plants". They are also called **dinoflagellates.**

Two members of the group, *Gymnodinium* sp. and *Gonyaulax* sp. are responsible for **"red tides"** in Florida and off the coast of California which color the water red and cause massive fish kills because of the toxic waste products they dump into the water during rapid growth periods, or **"blooms".**

A. Examine the slide of a **pyrrophyte** and locate the **transverse** and **longitudinal grooves** in which the two flagella lie. Also note the cellulose plates making up the cell wall.

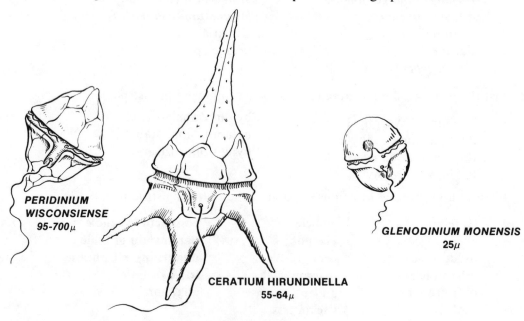

PERIDINIUM
WISCONSIENSE
95-700μ

GLENODINIUM MONENSIS
25μ

CERATIUM HIRUNDINELLA
55-64μ

FIGURE 15.1
REPRESENTATIVE DINOFLAGELLATES

IV. Phylum Chrysophyta — Golden Algae

Diatoms, by far the most numerous chrysophytes, are unicellular or colonial organisms of exceeding abundance in fresh and salt water, where they serve as food for small animals. Most of the plant portion of the **plankton** (floating life) are small algae like these, and by virtue of their vast numbers, their annual growth in terms of dry weight may equal all of the land plants together. They store oil instead of starch, and it is thought that much of our present undersea oil reserves may represent diatoms accumulated over millions of years, deposited with their shells on the ocean floor.

These organisms form cell walls composed largely of silica (the substance making up sand). The walls form two **valves** which fit together like the halves of a Petri dish. Diatoms which are oblong in shape are capable of smooth gliding movements on surfaces because of the presence of a **raphe,** or longitudinal slit, in their valves. Movement of the cell within the valves is thought to set up countercurrent water movement at the raphe, producing the gliding motion. Radially symmetrical forms lack raphes and thus are incapable of this type of movement.

A.   Examine a prepared slide of diatoms. Note the various shapes of cells and colonies. The following illustrations include some of the most commonly found diatoms.

B.   Make a fresh mount of pond or bog water. Notice the many types of diatoms. Some diatoms are colonial, while others are solitary forms.

C.   If time allows, make a slide of **diatomaceous earth.** This substance is an accumulation of diatom shells over a period of millions of years. Today, this soil is mined and used in silver polish, toothpaste, and water filters.

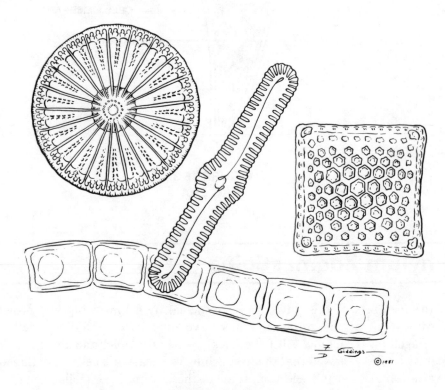

FIGURE 15.2
REPRESENTATIVE DIATOMS

117

V. Phylum Euglenophyta — Euglenoids

These protists occur mostly in fresh, but often polluted, waters. Their pigments are similar to those of higher plants but they neither store starch nor produce firm cellulose cell walls. Instead, euglenoids possess a flexible protein **pellicle** (inside the cell membrane) which allows the body to change shape while moving. Motility is achieved by an anterior whip-like **flagellum.**

A. Examine a prepared slide or make a fresh slide of *Euglena* and other euglenoids.

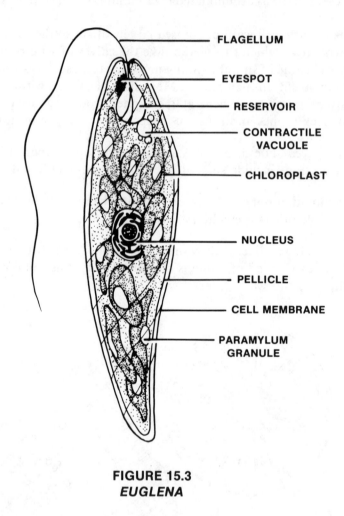

FLAGELLUM

EYESPOT

RESERVOIR

CONTRACTILE VACUOLE

CHLOROPLAST

NUCLEUS

PELLICLE

CELL MEMBRANE

PARAMYLUM GRANULE

FIGURE 15.3
EUGLENA

VI. Phylum Zoomastigina

All of the members of this phylum are propelled by a long whip-like projection called a **flagellum.** Almost all of the zoomastiginans have one, two, or more flagella; but in some members, a flagellum is lacking. All flagellates are heterotropic, and most are free-living, but some are parasitic. It is believed that the first primitive eucaryotic cells were flagellates. Because of genetic variation and natural selection, some of these primitive flagellates may have given rise to plantlike organisms while others may have evolved into more highly adapted and specialized animal types.

118

A. Obtain a slide of *Trypanosoma gambiense* and refer to Figure 15.4. This organism, carried by an intermediate invertebrate host, the **TseTse fly,** causes a neurological disease known as **African Sleeping Sickness** in man. Two morphological forms may be found on these slides; one with only an undulating membrane and one with an undulating membrane and flagellum.

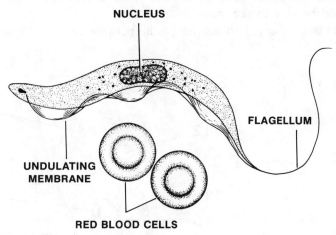

FIGURE 15.4
TRYPANOSOME IN BLOOD

VII.   Phylum Ciliophora

The organisms in this phylum are characterized by having hairlike projections called **cilia,** which they use for locomotion and food-getting. Cilia have the same basic structure as flagella, but are generally shorter and more numerous than flagella. Members of the phylum Ciliophora differ from other organisms in that they possess two different types of nuclei; a large **macronucleus,** and one or more smaller **micronuclei.** The macronucleus is concerned with the general control of cell activities, while the micronucleus is involved in sexual reproduction.

Like the flagellates, the ciliates are heterotrophic, a few being parasitic.

A.   Using the properly-labeled pipette in the bottle containing live specimens, prepare a wet mount of a *Paramecium caudatum.* In order to study the paramecium properly, you may have to slow it down. Add a drop of Protoslo to a drop of water containing some paramecia. Cover the preparation with a coverslip. Note: this organism is microscopic (invisible to the naked eye), but you can be relatively sure of getting a specimen if you will siphon water into the pipette from the material in the very bottom of the jar. The cilia cover the entire body and are relatively uniform in size. You may not be able to see the cilia with your microscope, but they should be visible in the demonstration slide under the phase contrast microscope. The basic shape of the animal resembles a slipper and its shape is maintained by its stiff outer covering, the **pellicle.** Note that one end is rounded and the other is pointed. In observing the living specimens, note that the same end is always directed anteriorly. Which end is it? Locate the **oral groove,** a depression into which food is swept by beating cilia. This leads to a tubular **gullet** where **food vacuoles** form. In some species undigestible wastes are released from the **anal pore,** which can only be seen when particles are being discharged through it. Two **contractile vacuoles** are present, one at each end of the cell. They contract alternately and function in expelling excess water. Just beneath the pellicle are located numerous **trichocysts** capable of being discharged like small darts under the right

119

conditions. Their function appears to differ in different ciliates, in some, functioning in defense, and in others in immobilizing prey. In paramecia, they appear to aid in adherence to surfaces while the organisms feed.

Apply a drop of weak acetic acid at the edge of the coverslip. As it proceeds under the coverslip and the paramecia come into contact with it, the trichocysts will discharge, and you should be able to see them.

Label the indicated structures of the paramecium on Figure 15.5.

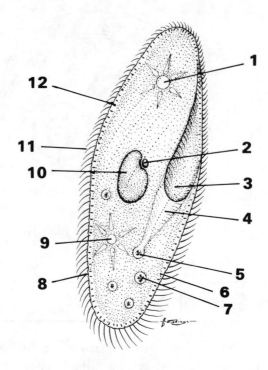

FIGURE 15.5
PARAMECIUM CAUDATUM

B. Occasionally you may observe the living paramecia reproducing. Ciliates may reproduce asexually by **binary fission;** dividing into two equal daughter cells. Since it is very difficult to see the nuclei in the living specimens, it is best to study a prepared slide. Examine prepared slides showing binary fission in *Paramecium* and sketch the various stages below.

C. In addition, Paramecia may reproduce sexually, allowing for greater genetic variability, through a process called **conjugation.** During this process, two individuals come to lie with their ventral sides together, and a protoplasmic bridge forms between them. The macronucleus disintegrates and gradually disappears. The micronuclei divide several times and finally the Paramecia exchange some of their nuclear material. This is a type of fertilization. Refer to your textbook for a detailed discussion of conjugation.

Examine a prepared slide showing *Paramecium* in conjugation and sketch several stages of conjugation.

CONJUGATION IN *PARAMECIUM*

D. Prepare a wet mount of pond water. Attempt to locate and identify some examples of Ciliophora. Refer to Figure 15.6 for identification and draw your beasties below.

POND WATER CILIOPHORA

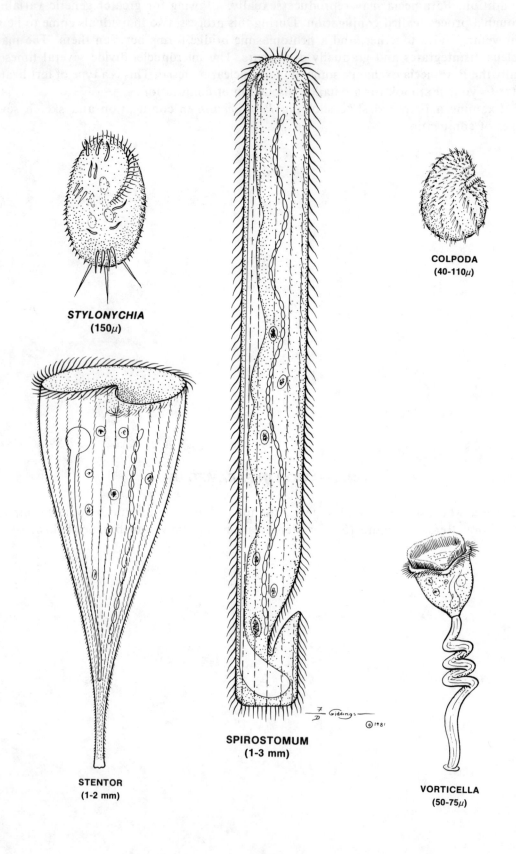

STYLONYCHIA
(150μ)

COLPODA
(40-110μ)

STENTOR
(1-2 mm)

SPIROSTOMUM
(1-3 mm)

VORTICELLA
(50-75μ)

FIGURE 15.6
REPRESENTATIVES OF THE PHYLUM CILIOPHORA

VIII. Phylum Sarcodina

The members of this phylum move by protoplasmic extensions called **pseudopodia** (false feet). The most common of these organisms is the *Amoeba,* which we will study in detail. The amoeba is a "blob of protoplasm" with no definite shape. Many other members of this phylum produce cells of varied shapes and compositions according to their species. All members of the phylum are heterotrophic.

A. Obtain a prepared, stained slide of *Amoeba proteus.* The organism is surrounded by a **plasma membrane.** Immediately inside this membrane is a thin, clear, non-granular layer of protoplasm called **ectoplasm.** Beneath this layer is the main body mass of granular **endoplasm.** Within the endoplasm is a large disclike structure, the **nucleus. Food vacuoles** may also be evident, containing partially digested granules of food taken in by the cell. Observe another vacuole towards the posterior end of the cell. This is the **contractile vacuole,** which functions to pump out excess water by alternately appearing as it fills up and disappearing as it empties.

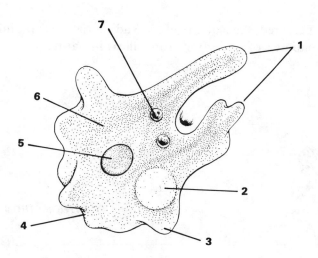

FIGURE 15.7
AMOEBA PROTEUS

B. Prepare a wet mount from the *Amoeba* culture. For the best results, reduce the amount of light passing through the specimen. Try to identify the anatomical structures of the *Amoeba.* Observe how the *Amoeba* moves by sending out its pseudopodia. Pseudopodia are also important in food-getting. They surround the food and enclose it in a food vacuole in the process known as **phagocytosis.** Also try to observe the pumping action of a contractile vacuole. Label the indicated structures in Figure 15.7.

C. Examine a prepared slide of the class Foraminifera. The organisms belonging to this class are mostly marine and have encompassing shells of various shapes. Each shell is composed of a **calcareous** material.

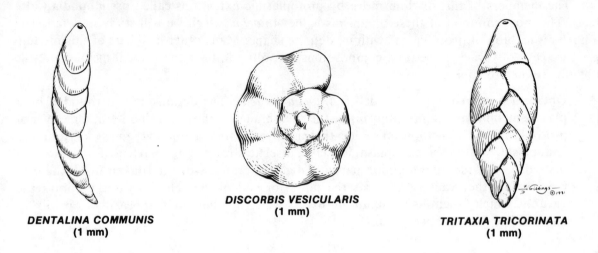

DENTALINA COMMUNIS
(1 mm)

DISCORBIS VESICULARIS
(1 mm)

TRITAXIA TRICORINATA
(1 mm)

FIGURE 15.8
REPRESENTATIVE FORAMINIFERANS

D. Examine a prepared slide of the class Radiolaria. The radiolarians have a shell that is siliceous (composed of or derived from **silica**) in nature.

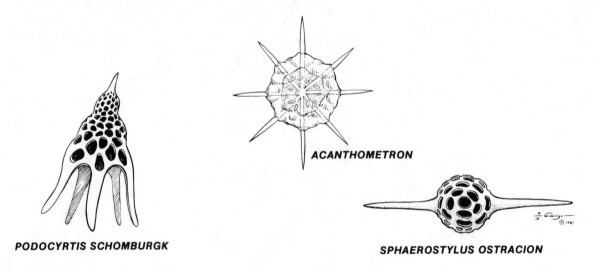

ACANTHOMETRON

PODOCYRTIS SCHOMBURGK

SPHAEROSTYLUS OSTRACION

FIGURE 15.9
REPRESENTATIVE RADIOLARIANS

IX. Phylum Sporozoa

Some sporozoans have pseudopods and some have flagellated gametes. The most widely-known member of the sporozoan phylum is *Plasmodium*, the blood parasite which causes malaria in man. Like *Trypanosoma, Plasmodium* is carried by a secondary host, the female *Anopheles* mosquito, which transmits malaria to man. All of the sporozoans are highly specialized in particular parasitic ways of life. They typically have a life cycle in which a sexual stage alternates with a spore-producing asexual stage. Spores develop as the result of multiple fission of a "mother cell." Some are amoeboid and others are encapsulated.

A.   Examine a slide of *Plasmodium malariae* and sketch your observation below.

PLASMODIUM MALARIAE

X.   Review Questions

A.   *Trypanosoma gambiense* causes _____ in man.

B.   By what means does *Paramecium* move? _____

C.   _____ move throughout the cytoplasm in a regular pattern and finally discharge all undigested material through the anal pore of the paramecium.

D.   Paramecia may reproduce asexually by means of _____ and sexually by means of _____ .

E.   The amoeba moves by _____ which means " _____ ".

F.   Why would you expect a freshwater amoeba to be more likely to have contractile vacuoles than a marine amoeba? _____

_____

G.   *Plasmodium malariae* causes _____ in _____ .

H.   Which protists are heterotrophic: euglenoids, chrysophytes, or protozoans? _____

_____

I   The cell wall of diatoms is largely composed of what durable material? _____

_____

J.   What is the storage product of diatoms? _____

K.   How do diatoms differ from euglenoids in their movement? _____

_____

L.   How does a pellicle differ from a cell wall? _____
_____

M.   Planktonic organisms are _____
_____

N.   What is a "red tide"? _____

KINGDOM ANIMALIA:
PORIFERA, CNIDARIA, AND CTENOPHORA PHYLA

I. Objectives

After the completion of this exercise, the student should be able to do each of the following:

A.   Compare the Phyla Porifera, Cnidaria, and Ctenophora as to their level of organization (cellular, tissue, organ, system).

B.   Describe the feeding and nutrient processing of a sponge and a hydra.

C.   Identify the anatomical structures of the sponge and the hydra.

D.   Define the terms given in boldface other than the anatomical structures.

E.   Identify the phyla and classes of each of the animals in the jars on display.

F.   List the basic cell types in the sponge and the function of each.

G.   Compare and contrast the two body forms of the Cnidaria.

H.   Describe the alternation of asexual and sexual reproduction shown by *Obelia*.

I.   Answer the review questions at the end of this exercise.

II. The Animal Kingdom

The animal kingdom is composed of multicellular, obviously motile, heterotrophic organisms which reproduce sexually by oogamy and produce multicellular embryos. You may wish to refer to Exercise 6 and review the outline of animal taxa to be studied in the next several exercises.

III. Phylum Porifera

The phylum Porifera contains animals known as sponges. Porifera means "bearing pores". Sponges are multicellular, but the cell aggregates of this phylum do not form true tissues. However, differentiation of cells has occurred so that there is a division of labor such as reproduction, feeding, and water circulation.

Many of these animals have body shapes resembling vases (see Figure 16.1). The walls of sponges have numerous tiny openings called **ostia,** through which currents of water enter carrying food and oxygen to the central cavity or **spongocoel.** The movement of water is produced by the beating of the flagella of **collar cells,** or **choanocytes,** in the **flagellated chambers,** or **radial canals.** The choanocytes capture and digest food particles brought in by the water currents with the aid of collar-like structures surrounding their flagella. The water entering the ostia circulates through the **incurrent canal, prosopyles** (openings between the incurrent canals and radial canals), **radial canal, apopyles** (openings between the radial canals and the spongocoel), and **spongocoel** and exits through the opening at the top of the animal. This opening is called the **osculum.**

The body wall consists of three layers. The outer layer consists of flat **epithelial cells,** among which are contractile cells called **pinacocytes** which regulate the sizes of the ostia. The middle layer consists of gelatinous non-living matrix containing living mesenchyme cells called **amoebocytes,** or **archeocytes,** which are capable of amoeboid movement. Amoebocytes have many functions. They collect food from the flagellated collar cells, secrete the gelatinous matrix, collect wastes, produce **spicules,** and can differentiate into any of the other cell types. The spicules, which are minute crystal-like structures composed of calcium salts or silicious material, form the supportive skeleton of the sponge.

Sponges are unique in that it is believed that they have evolved from a completely different group of flagellates than did other animals. For this reason, they are considered to be an evolutionary dead end.

Reproduction in sponges occurs asexually by **budding, fragmentation,** and in freshwater forms, by **gemmule** formation. A gemmule consists of a ball of amoebocytes surrounded by a capsule consisting of spicules and dead cells. Sexual reproduction in sponges involves the fusion of eggs and sperm formed from amoebocytes. The zygote develops into a free-swimming ciliated larva.

Obtain a prepared slide of the cross section of *Grantia.* Make a sketch of what you see, labelling the central cavity, or spongocoel, incurrent canal, radial canal, and choanocytes. Refer to Figure 16.1. Also note the small openings, or apopyles, leading from the radial canals into the spongocoel.

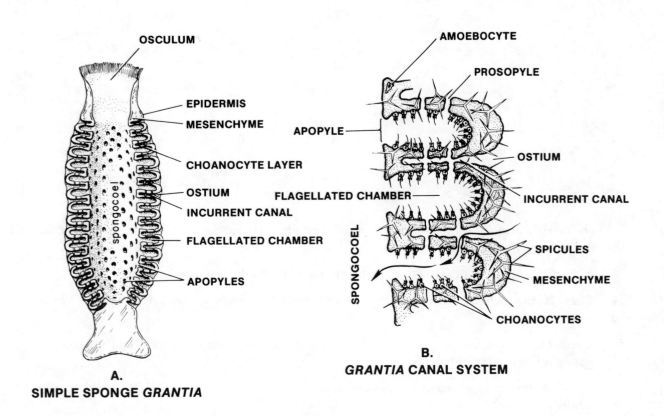

OSCULUM

EPIDERMIS
MESENCHYME

CHOANOCYTE LAYER

OSTIUM
INCURRENT CANAL

FLAGELLATED CHAMBER

APOPYLES

spongocoel

A.
SIMPLE SPONGE *GRANTIA*

AMOEBOCYTE

PROSOPYLE

APOPYLE

OSTIUM

FLAGELLATED CHAMBER

INCURRENT CANAL

SPICULES

MESENCHYME

CHOANOCYTES

SPONGOCOEL

B.
GRANTIA CANAL SYSTEM

FIGURE 16.1
GRANTIA (L.S.)

CROSS SECTION OF GRANTIA

Obtain another slide showing the **spicules** which form the skeleton of the sponge. Make a sketch of *Grantia* spicules below.

GRANTIA SPICULES

The classification of sponges is determined largely by shape and chemical composition of the skeleton. The phylum Porifera is subdivided into three classes:

A.   Class Calcarea — Calcareous or chalky sponges. Example — *Grantia*

B.   Class Hexactinellida — Glass sponges, composed of siliceous spicules. Example — Venus Flower Basket

C.   Class Demospongiae — Commercial or bath sponges. Skeleton includes proteinaceous **spongin** fibers. Example — *Spongia*.

Examine the specimens of sponges in the display jars, noting the class to which each belongs.

IV.   Phylum Cnidaria (Coelenterates)

The animals of this phylum demonstrate the tissue level of organization. Although the cell aggregates function as tissues, no true organs are present. The body of the animal is composed of two epithelial layers: an outer **epidermis** and an inner **gastrodermis,** which lines the **gastrovascular cavity,** or **coelenteron.** Between these two layers is a gelatinous layer of **mesoglea.** The epidermis is a well developed layer of closely-packed cells which function to protect the organism and obtain food. The gastrodermis is also well developed and serves in digestion and internal transport. The mesoglea is poorly developed, varying from a jellylike substance containing a few cells which serve in coordinating the actions of the organism and the production of gametes, to a true cellular layer in the most advanced members of the group.

The most distinguishing characteristic of this phylum is the possession of specialized cells called **cnidoblasts.** These cells contain stinging structures called **nematocysts** which are used for defense and food capture. A nematocyst is a capsule containing a long coiled thread which shoots out and either traps and holds the prey or injects a toxic substance which paralyzes the prey or predator.

Another characteristic of these animals is that they are polymorphic, displaying different body forms at different points in the life cycle of the organism. There are two basic body forms; the **polyp** and the **medusa.** The medusa is generally a more active swimming form where as most polyps are **sessile,** remaining attached to some substrate. Both forms are found in the life cycle of many coelenterates. Some coelenterate colonies may be composed of both forms of individuals. Both forms are basically **radially symmetrical.**

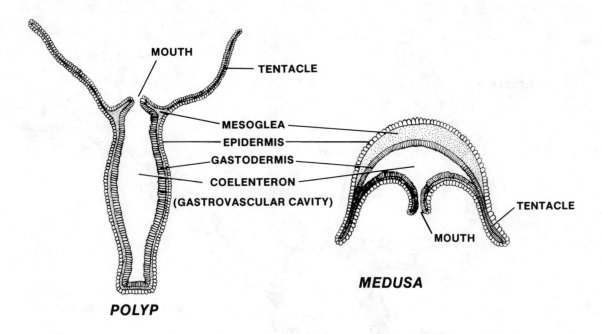

FIGURE 16.2
BODY FORMS OF CNIDARIA

The phylum Cnidaria is subdivided into three classes:

A. Class Hydrozoa

Many members of this class develop into sessile polyp colonies such as *Obelia*. Their life cycles characteristically involve a regular alternation between asexual (polyp form) and sexual (medusa form) reproduction, but the polyp form is generally dominant.

Hydra is an exceptional member of this class. It is a mobile individual polyp rather than a sessile or colonial form. Reproduction occurs asexually by budding or sexually by the production of sperm and eggs, and only polyps are produced. There are no medusae produced at any time.

Obtain a prepared slide of a *Hydra*. Note the following parts (Refer to Figure 16.3) as you find them; the **basal disc,** for attachment, the cylindrical **body,** the circle of **tentacles,** the elevated **hypostome** at the base of the tentacles, a **mouth** in the center of the **hypostome, buds, spermaries** (swellings just beneath the tentacles), **ovaries** (swellings in the lower portion of the body), **nematocysts, gastrovascular cavity, epidermis, mesoglea, gastrodermis,** and **flagellum.**

Prepare a wet mount of living *Hydra* using a concave depression slide. **Do not cover with a cover slip.** For the best results, reduce the amount of light entering the slide. Do not jar the table or microscope, as any disturbance will cause the *Hydra* to contract. Feeding the *Hydra* will be optional. After viewing the living *Hydra,* return it to the container marked **Fed Hydra.**

Obtain a prepared slide of the colonial *Obelia* and find the parts listed on Figure 16.4. This colony forms from the repeated budding of a single individual. The various buds take up either reproductive (**gonangium** or reproductive polyp) or feeding (**hydranth** or feeding polyp) duties that are performed for the entire colony. The reproductive polyp, or gonagium, will produce medusae which will in turn produce eggs or sperm. When fertilized, or when egg and sperm unite, the zygote will go through various stages of embryonic development and will give rise to a ciliated larvae, the **planula.** The planula attaches itself to the substratum and gives rise to a new asexual hydroid colony.

131

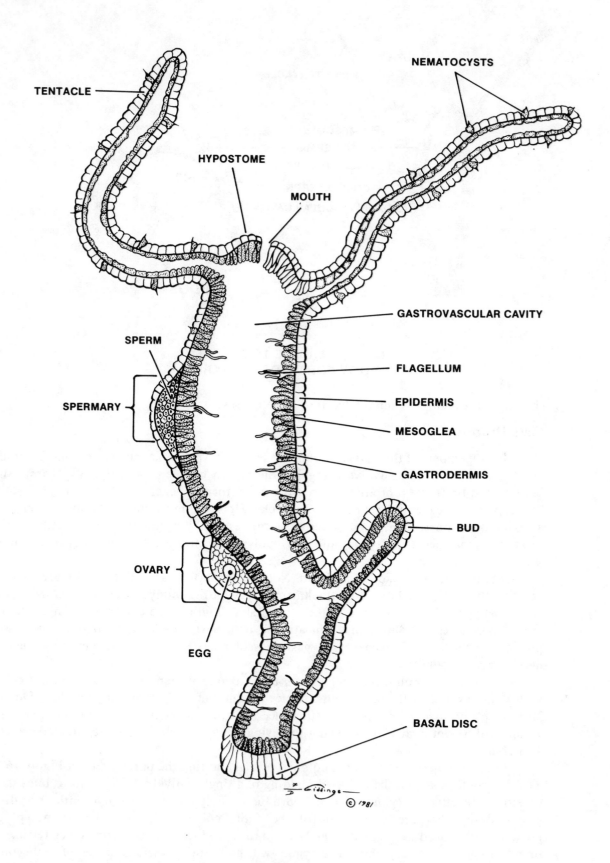

TENTACLE

NEMATOCYSTS

HYPOSTOME

MOUTH

GASTROVASCULAR CAVITY

SPERM

FLAGELLUM

SPERMARY

EPIDERMIS

MESOGLEA

GASTRODERMIS

BUD

OVARY

EGG

BASAL DISC

© 1981

FIGURE 16.3
HYDRA (L.S.)

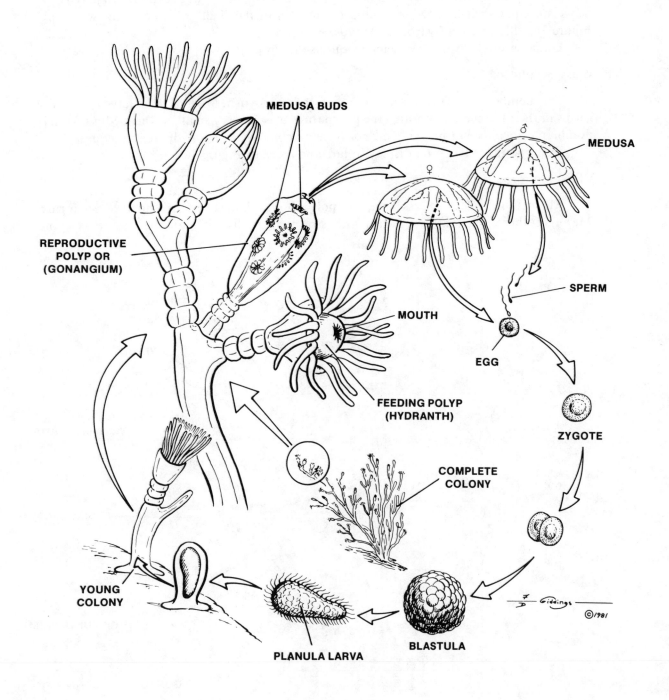

MEDUSA BUDS

MEDUSA

REPRODUCTIVE
POLYP OR
(GONANGIUM)

SPERM

MOUTH

EGG

ZYGOTE

FEEDING POLYP
(HYDRANTH)

COMPLETE
COLONY

YOUNG
COLONY

PLANULA LARVA

BLASTULA

FIGURE 16.4
OBELIA **HYDROID COLONY**

133

Another example of a Hydrozoan is *Gonionemus,* which has a dominant medusa stage. For years, *Gonionemus* was thought to lack the polyp stage because it was so small (1 mm.) that no one had been able to detect it. The polyp can not only produce medusae by budding, but can also bud to produce other polyps. Upon the completion of its development, the medusa form of *Gonionemus* resembles a typical jellyfish, but unlike scyphozoan jellyfish, it has a muscular shelf-like **velum** around the margin of the "bell" which aids in swimming. Figure 16.5 illustrates a hydrozoan medusa.

Examine the hydrozoan specimens on display in jars.

B.   Class Scyphozoa

The members of this class are commonly called **jellyfish.** Most of them have both body forms in their life cycles, although the polyp form may be very small, as the medusa form is dominant. Scyphozoan medusae do not have a velum as do the hydrozoan medusae.

Examine the jellyfish on display in jars and refer to Figure 16.6.

C.   Class Anthozoa

All members of this class have the polyp body plan. There are no medusa forms. Many of them, such as the corals and sea pansies, live in colonies, whereas some

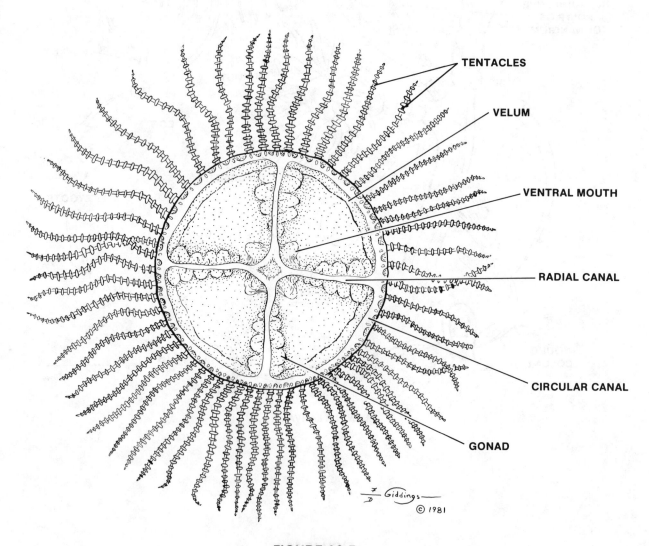

FIGURE 16.5
GONIONEMUS MEDUSA

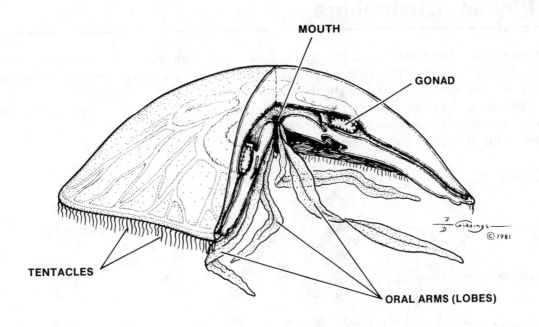

MOUTH

GONAD

TENTACLES

ORAL ARMS (LOBES)

FIGURE 16.6
STRUCTURE OF JELLYFISH, AURELIA

such as the sea anemone, live independently. Examine the representatives of this class on display and compare them with the diagrams below.

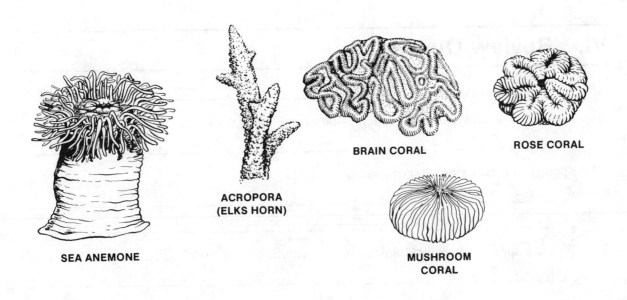

SEA ANEMONE

ACROPORA
(ELKS HORN)

BRAIN CORAL

ROSE CORAL

MUSHROOM
CORAL

FIGURE 16.7
REPRESENTATIVE ANTHOZOANS

135

V. Phylum Ctenophora

Members of phylum Ctenophora are often known as the **comb jellies** and **sea walnuts.** They comprise about 80 species of free-swimming marine animals with translucent gelatinous bodies. Ctenophores show resemblance to jellyfishes and at one time were classified with the coelenterates. The unique characteristic is that they possess eight rows of swimming "combs", or **ctenes,** which are composed of fused cilia; unlike the cnidaria, they lack nematocysts.

Observe the ctenophores on display.

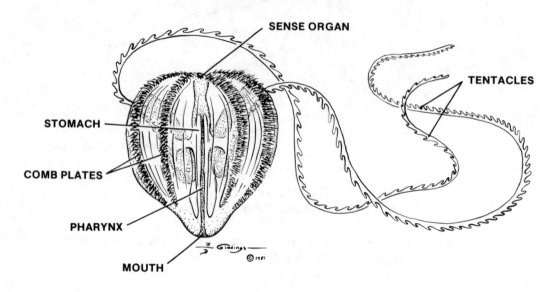

FIGURE 16.8
CTENOPHORE

VI. Review Questions

A. All members of the Porifera exhibit the _multicellular, division of labor_ level of organization.

B. Two kinds of cells showing division of labor in a sponge are _____
and _____ .

C. Sponges reproduce by (describe each):

   1. _budding, fragmenting_ — _involves fusion of eggs & sperm formed from amoebocytes_

   2. _asexually fragmenting ball of_ (fresh water) _by amebocytes surrounded by specialist & dead cells_

D. Why are sponges considered unique in the evolution of animals? _____
   They are a dead end species evolved from a completely different flagellate than other animals

E. Organisms in the Phylum Cnidaria represent the ___tissue___ level of organization.

136

F. Identify the following:
1. gastrovascular cavity _serves in digestion + in nu transport_
2. nematocyst _stinging structures - for defense for food · capsule threaded_
3. planula _ciliated larvae_
4. velum _(unlike jellyfish) has muscular shelf which aids in swimming_
5. polyp _sessile - attaches to substrate - no movement - floats_
6. medusa _active swimming_
7. archeocyte _gelatinous non-living matrix mesenchyme containing living cells_
8. spicule _crystal-like spikes - calcium salts or silicas - skeleton of sponge_
9. choanocyte _beating the flagella of causes movement of water_

G. The unique characteristic of the Ctenophores is _8 rows of swimming combs_

KINGDOM ANIMALIA:
PLATYHELMINTHES, NEMATODA, ROTIFERA, AND ANNELIDA PHYLA

exercise 17

I. Objectives

After the completion of this exercise, the student should be able to do each of the following:

A. Identify the phylum and class of each of the animals in the jars on display.

B. Define the boldface terms referring to the earthworm.

C. Describe two evolutionary advantages possessed by members of both the phylum Nematoda and the phylum Rotifera.

D. Examine the *Ascaris* cross section slide and identify its component parts.

E. Distinguish between a male and female nematode.

F. Examine the rotifer slide and identify the "wheels" and the "forked foot."

G. List, cite an example of, and identify the major characteristics of each of the three major classes of the phylum Annelida.

H. Dissect an earthworm and identify the boldface anatomical parts referred to in the directions.

I. Identify the structures referring to the earthworm cross section slide printed in boldface in the text.

J. Distinguish between a **pseudocoelom** and a true **coelom** and explain the advantage of possessing a true coelom.

K. List the advantages of the members of the phylum Platyhelminthes over the members of the phylum Porifera and phylum Cnidaria.

L. Identify the anatomical structures of planarians, flukes, and tapeworms given in boldface in the text.

M. List, cite an example of, and identify the major characteristics of each of the three classes of the phylum Platyhelminthes.

N. Answer the review questions at the end of this exercise.

II. Phylum Platyhelminthes

Members of this phylum are commonly referred to as the flatworms. As the name implies, they are flattened dorsoventrally. The flatworms show many advantages over the Porifera and Cnidaria, such as:

1. Bilateral symmetry.

2. A complex organ-system level of organization.

3. A mesodermal germ layer. They are therefore referred to as being **triploblastic** in general structure.

4. A central nervous system showing **cephalization.**

5. A distinct head with sense organs.

Other characteristics of this phylum include the absence of a body cavity, that is, they are **acoelomate;** the absence of an anus; and the combining of sexes within single animals **(hermaphrodism).** There are both parasitic and free-living forms.

The phylum Platyhelminthes is subdivided into three classes:

A. Class Turbellaria

This class includes the free-living flatworms, an example of which is *Dugesia,* the common planaria. Members of this class may be found under rocks or attached to submerged objects in the clear water of lakes, springs, and streams.

Obtain and examine a prepared slide of the freshwater planaria. In your examination, note the general body shape, and give special attention to the head, **eyespots,** and **auricles** (lateral projections on the head which function as tactile and chemosensory organs in the anterior region). On this slide the digestive system is stained by feeding the worms India ink before preserving them; it clearly demonstrates the branching of the **gastrovascular cavity.** Identify the muscular **pharynx,** or **proboscis,** which is withdrawn into the **pharyngeal pouch** in the middle area of the body. This pharynx may be extended out of the body while feeding.

Prepare a wet mount of living planaria using a concave depression slide. **Cover with a cover slip.** Notice the general shape and the mode of locomotion of the flatworm. Feeding the planaria with bits of liver is optional. After viewing the planaria, return them to the container marked **Fed Planaria.** Also examine the preserved specimens on display in jars.

B. Class Trematoda

This class is composed of parasitic flatworms known as the flukes. They have evolved a thick, protective outer layer of nonliving substance known as the **cuticle** which is secreted by the epidermis and protects the organism from being digested by the host's digestive enzymes.

Examine the slide of *Clonorchis sinensis* (whole mount) under low power. *Clonorchis,* the Chinese liver fluke, possesses an anterior **oral sucker** and a large **ventral sucker.** Beginning anteriorly at the oral sucker, trace the digestive tract to the short muscular **pharynx,** and then to the forked **gastrovascular cavity.** The reproductive system includes a mass of **testes** located at the posterior end. Sperm that is produced here moves through a duct to the **genital pore** located just anterior to the ventral sucker. Eggs are produced in

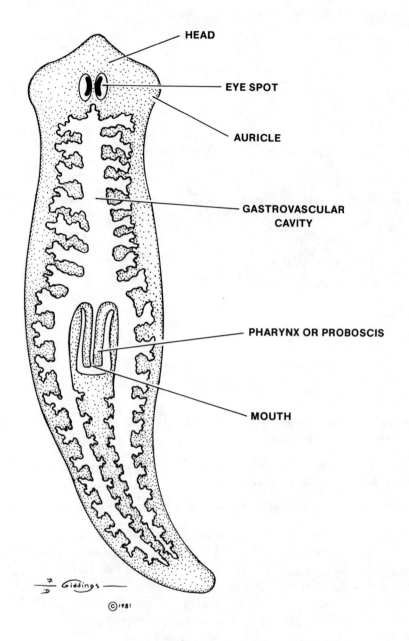

HEAD

EYE SPOT

AURICLE

GASTROVASCULAR
CAVITY

PHARYNX OR PROBOSCIS

MOUTH

FIGURE 17.1
PLANARIA: DIGESTIVE SYSTEM

an **ovary,** anterior to the testes. In front of the ovary is a large convoluted duct called the **uterus** which serves as a storage area for fertilized eggs.

Label the underlined anatomical structures of the "Chinese liver fluke" on Figure 17.2. Also examine the preserved specimens of trematodes in jars.

C. Class Cestoda

This class includes the tapeworms, all of which are parasitic. Examine the preserved specimens on display and obtain a prepared slide of *Taenia pisiformis,* the dog tapeworm. This organism is transmitted by ingestion of infected fleas and may also infect humans in this way.

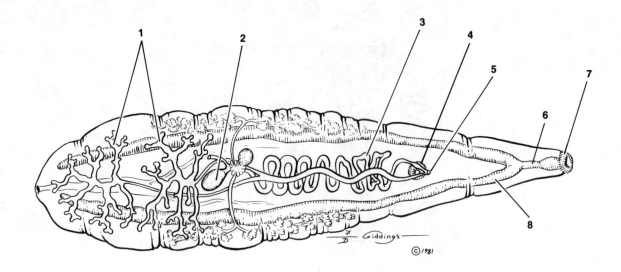

FIGURE 17.2
CLONORCHIS SINENSIS

Examine the stained slide and, with the help of Figure 17.3, identify the following areas and their associated structures.

1. **Scolex,** or head with **rostellum,** which attaches to the intestinal wall of the host, has rows of hooks and four **suckers.**

2. **Neck,** has no segmentation and represents a growing area.

3. The body, or **strobila,** consisting of units called **proglottids.**

    The proglottids behind the neck are young, or **immature** and only contain the male sex organs while proglottids in the middle region are **mature** and contain both male and female sex organs. At the posterior end, the proglottids lack male reproductive organs as they have disintegrated and the proglottids in this region are **gravid,** or "ripe" sections filled with fertilized eggs which will become detached and pass out of the host via the feces, to be picked up by another successive host.

III. Phylum Nematoda

This phylum includes the unsegmented roundworms. Some are parasitic, but most are free-living. An advancement of this phylum over the ones previously studied is that animals in this phylum possess a **complete digestive tract** which has two openings, a mouth and an anus, allowing for oneway passage of ingested food. A second advancement is the presence of a body cavity. This cavity is not a true **coelom** since it does not lie between layers of mesoderm. The cavity, called a **pseudocoelom,** is found between the outer body wall and the digestive tube. A unique characteristic of this group is that each animal has a limited number of cells in its body. This specific number of cells is characteristic of its species. Animal growth in this phylum beyond the

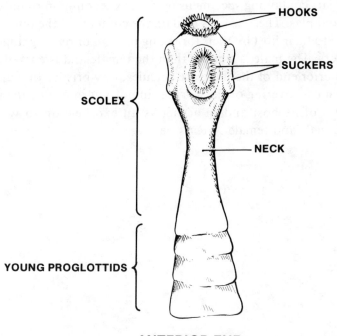

HOOKS

SUCKERS

SCOLEX

NECK

YOUNG PROGLOTTIDS

ANTERIOR END

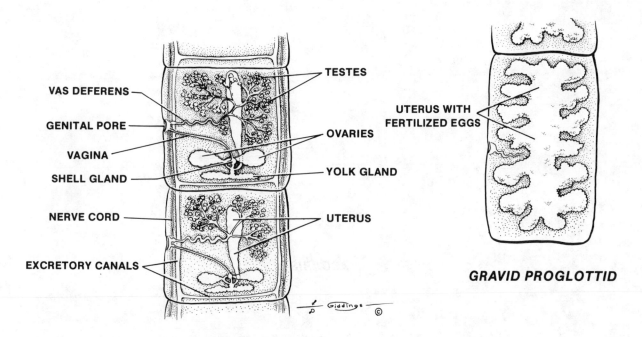

TESTES

VAS DEFERENS

GENITAL PORE

VAGINA

OVARIES

SHELL GLAND

YOLK GLAND

NERVE CORD

UTERUS

EXCRETORY CANALS

UTERUS WITH
FERTILIZED EGGS

Giddings ©

GRAVID PROGLOTTID

MATURE PROGLOTTIDS

FIGURE 17.3
TAENIA PISIFORMIS

embryo stage is due to cell growth rather than cell multiplication as in other phyla. Their musculature consists mainly of longitudinal muscles which cause them to move in a whip-like fashion.

A. Examine the display jars containing specimens of *Ascaris*, a common roundworm found as a parasite in humans and pigs. These worms feed on the contents of the intestinal tract. Some damage may be done to their host by their thrashing motion or by clogging of the intestinal tract by large numbers. The male is smaller than the female and is curved at the posterior end. Also at the posterior end of the male are spicules, tiny bristles that aid in copulation. The **cuticle,** or outermost covering of the worm is important in protecting the animal from the digestive enzymes of its host and in acting as an exoskeleton to which muscles are attached. Sketch the male and female *Ascaris* below.

ASCARIS

B. Examine the *Ascaris* cross section slide. Refer to Figure 17.4 and identify the labeled parts in the figure.

C. Examine slides of *Ancylostoma*, a human hookworm. The anterior end of the hookworm has hooks with which it attaches to the intestinal wall of the host. Unlike *Ascaris*, the hookworm feeds on its host's blood. You can distinguish the male from the female by

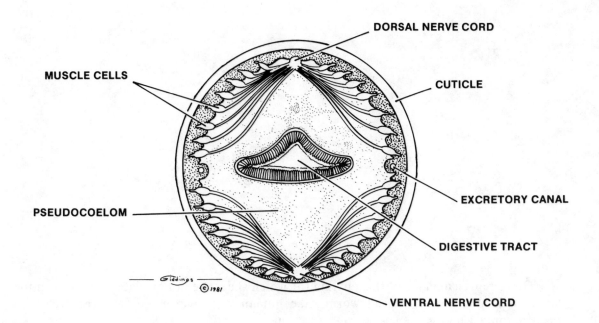

FIGURE 17.4
ASCARIS **(X.S.)**

looking at the posterior end of the worm, which in the male is modified for copulation by way of a fanlike structure called a **bursa**. Sketch the male and female hookworms indicating the anterior and posterior ends.

MALE HOOKWORM   **FEMALE HOOKWORM**

D. Examine a slide of the voluntary muscle fibers of rat or man containing specimens of encysted *Trichinella*. This roundworm forms cysts in the muscle of hogs and may infect humans who eat insufficiently cooked pork. The cyst you observe on the slide is only one stage in the life cycle of this animal. Sketch and label: **muscle tissue of host, cyst wall,** and **worm.**

Other representatives of this phylum for which slides are not available for study in the lab include pin worms, filarial worms, the human whip worm, and vinegar "eels", to mention only a few.

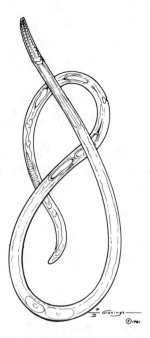

FIGURE 17.5
FREE-LIVING NEMATODE

E. Examine pond water for free-living nematodes. You should be able to identify them by their whip-like thrashing motion (refer to Figure 17.5). Attempt to isolate one on a slide and examine it more closely. Make a brief sketch of the animal on the next page.

IV. Phylum Rotifera

The rotifers or "wheel-bearing animalcules" are usually less than 1 mm. in length and most are found in freshwater. Most of the 1,500 species of Rotifers are free-living, a few are parasites, and a few inhabit salt water. Members of this phylum possess both a complete digestive tract and a **pseudocoelom** (pseudo — false; coelom — body cavity). The distinguishing characteristics of the animals included in this phylum are:

1. "Wheels" of beating cilia on the anterior end of the body.

2. Absence of external cilia elsewhere.

3. A chewing **pharynx** or **mastax** which is used for grinding the ingested food particles.

4. "Forked foot" on the posterior end of the body.

5. Growth beyond the embryo stage is by cell growth rather than by cell multiplication.

Examine a slide of a Rotifer whole mount. Identify the **"wheels"** and the **"forked foot."** Use Figure 17.6 to help you.

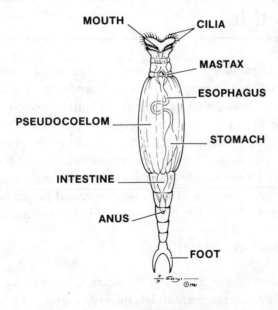

FIGURE 17.6
ROTIFER

A.   Examine pond water for aquatic rotifers. If they are moving quickly, add a drop of Protoslo to the slide under the cover slip. Make sketches of the rotifers observed.

ROTIFERS

V.   Phylum Annelida

This phylum consists of approximately 7,000 species of **segmented** worms which are divided among 4 classes: **Polychaeta,** the marine bristleworms and sandworms; **Hirudinea,** the leeches; **Oligochaeta,** the earthworms; and **Archiannelida,** a small, primitive group of tiny marine worms. Study the specimen jars on display for the first three classes.

A unique feature of the phylum Annelida not previously seen in the organisms studied, is the presence of a true body cavity or **coelom.** The coelom is completely lined by mesodermal tissue. With a true body cavity there is an independence of the digestive tract from the muscles of the body wall. Therefore, rhythmic contraction of these muscles, called **peristalsis,** allows food to be moved through the digestive tract without the movement of the entire animal.

A.   Class Polychaeta (many bristles)

The polychaeta represents the largest class of annelids. They possess fleshy tentacles on the head, and two fleshy appendages, **parapodia,** on each segment except the first and last. Many **setae,** or chitinous bristles, are found on the parapodia which are used for swimming, burrowing, crawling, and gas exchange. See figure 17.7.

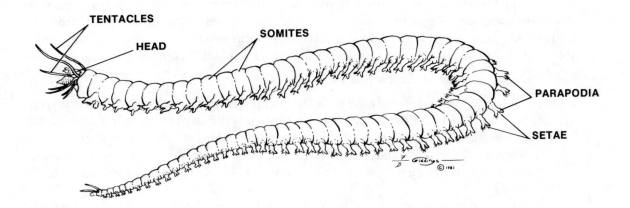

FIGURE 17.7
CLAMWORM

B.   Class Hirudinea

   The leeches lack tentacles, parapodia, and setae. Many of the members of this class are parasitic and inhabit freshwater. An interesting feature of the leeches is that they possess two muscular suckers: small anterior suckers which surround their mouths and large posterior suckers used in locomotion and attachment.
   Examine the slide of a leech and locate the structures indicated below.

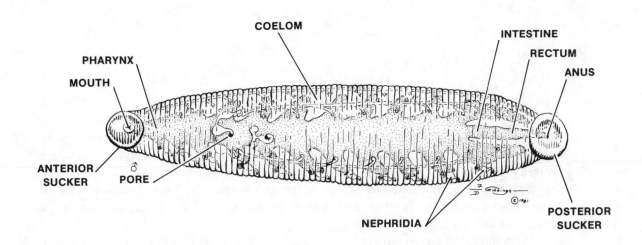

FIGURE 17.8
LEECH

C. Class Oligochaeta (few bristles)

Members of this class inhabit damp soil and freshwater. They are somewhat degenerate annelid types with the head and locomotor structures greatly reduced. They lack parapodia, having only setae.

The earthworm will be studied extensively in lab. Before you start dissecting, **read the directions carefully.** Remember that the diagrams are intended to help you find the anatomical parts, but you may ask your instructor for assistance. It is essential that you be able to identify the parts in the dissected animal and not just memorize the figures.

1. External Anatomy of the Earthworm

Obtain a specimen of *Lumbricus* (the "nightcrawler" earthworm) and run your fingers over the surface of its body. Do you note any differences between the resistance to the motions of your fingers in different directions? This is caused by the presence of the **setae** on the ventral and ventrolateral surfaces. Place the worm on moist paper towelling in a dissecting pan. Refer to Figures 17.9, 17.10 and 17.11 as you perform the dissection.

The most obvious feature of the earthworm is its segmentation. The very small first segment overhangs the mouth and is known as the **prostomium** (meaning in front of the mouth). Note that it does not have a corresponding ventral portion. Beginning with the

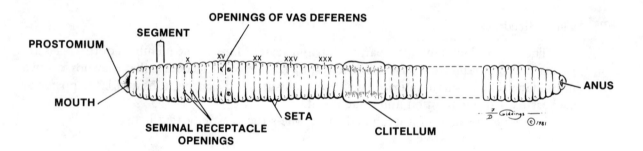

FIGURE 17.9
EARTHWORM: VENTRAL VIEW

next segment (the first complete one) we will assign numbers to the segments for convenience. Several segments, beginning with number 32 or 33 are swollen because of large hypodermal glands responsible for the formation of the cocoon. These swollen segments comprise the **clitellum** which is located anteriorly. This structure is not as obvious from the ventral surface as it is from the dorsal surface. The anus is located at the end of the last segment.

With the aid of a stereomicroscope, determine the number and location of the setae and their orientation on a segment. Locate the openings of the sperm ducts or vasa deferentia which lie ventrolaterally on segment 15; they are a pair of transverse slits lying between two swollen lips. The oviduct openings are similarly located on segment 14, but are less conspicuous.

150

Place the worm, ventral side down, in a dissecting pan. Carefully pin your specimen with pins through the prostomium and the posterior segment. Make a short, longitudinal incision in the dorsal midline, forward through the body wall from a short distance back of the clitellum to the anterior end. Be very careful not to cut through more than just the body wall, noting the **septa** (thin membranes). Pin the body flat by placing the pins in every fifth segment, and lean each pin toward the outer edges of the pan so that your view will be unobstructed.

2. Internal Anatomy

   a. Circulatory System

   The earthworm has a circulatory system consisting of five pairs of **hearts** surrounding the **esophagus,** one pair each in segments 7 through 11. The **dorsal blood vessel** located along the middorsal line above the digestive tract, carries blood posteriorly. If you cannot see it at this stage, do not cut away to find it; search later. Smaller blood vessels will be seen on the outer surface of the gut and the inner surface of the body wall.

   b. Digestive System

   Identify the **mouth** and the **buccal cavity** which extend through the first three segments. More conspicuous is the *pharynx,* a thick muscular organ with accessory lubricating glands inside, occupying segments 3 to 5. From this segment to segment 14 is the relatively slim **esophagus.** In segments 15 through 17, the digestive tract expands into the **crop** where food is stored. Posterior to the crop is a thick-walled **gizzard,** a mastication organ. The **intestine** is located beyond the gizzard and leads to the **anus.** Most digestion and absorption takes place in the intestine. It is well supplied with secretory cells. The surface area of the intestine through which absorption can take place is greatly increased by two devices: segmental constrictions and the **typhlosole.** The typhlosole is an internal longitudinal ridge of the intestinal wall. Use Figure 17.10 in order to help you find the typhlosole.

   c. Excretory System

   Each segment, except for the first three and the last, has a pair of white tubular excretory organs, the **nephridia,** which lie lateral to the gut. Each organ opens to the outside through its own duct and pore. These pores are difficult to see.

   d. Nervous System

   Extend the middorsal incision from the clitellum towards the anus. Remove the intestine carefully to expose the **ventral nerve cord** located beneath the **ventral blood vessel. Ganglia** (singular, ganglion) are present along the cord in each segment and handle much of the coordination of these animals without the intervention of the main brain. The nerve cord and the ganglia are difficult to see in worms. You should see (unless you accidentally cut it away) the **brain,** composed of a pair of white ganglia above the pharynx in segment 3. These communicate with the ventral nerve cord through a pair of **circumpharyngeal connectives.**

   e. Reproductive System

   The most obvious portions of the male reproductive tract are the two three-lobed **seminal vesicles,** or sperm reservoirs, usually cream-white in color. Fastened ventrally and extending dorsally around each side of the esophagus, they include the two small **testes** within them. Sperm are freed from the testes and complete their

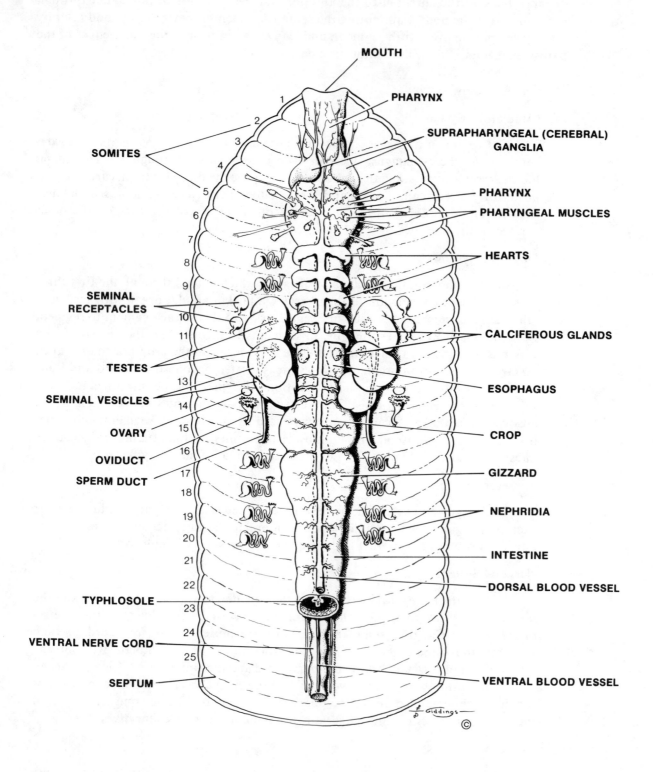

MOUTH

PHARYNX

SUPRAPHARYNGEAL (CEREBRAL) GANGLIA

PHARYNX

PHARYNGEAL MUSCLES

HEARTS

CALCIFEROUS GLANDS

ESOPHAGUS

CROP

GIZZARD

NEPHRIDIA

INTESTINE

DORSAL BLOOD VESSEL

VENTRAL BLOOD VESSEL

SOMITES

SEMINAL RECEPTACLES

TESTES

SEMINAL VESICLES

OVARY

OVIDUCT

SPERM DUCT

TYPHLOSOLE

VENTRAL NERVE CORD

SEPTUM

FIGURE 17.10
EARTHWORM: DORSAL DISSECTION

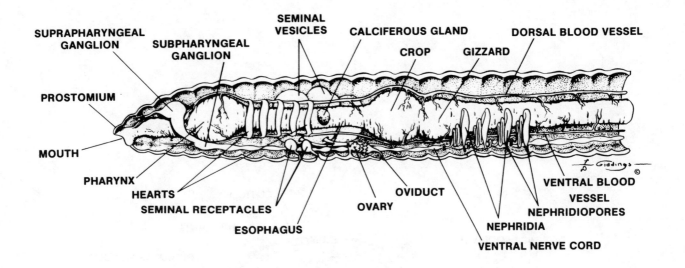

SUPRAPHARYNGEAL GANGLION
SUBPHARYNGEAL GANGLION
SEMINAL VESICLES
CALCIFEROUS GLAND
CROP
GIZZARD
DORSAL BLOOD VESSEL
PROSTOMIUM
MOUTH
PHARYNX
HEARTS
SEMINAL RECEPTACLES
ESOPHAGUS
OVARY
OVIDUCT
VENTRAL BLOOD VESSEL
NEPHRIDIOPORES
NEPHRIDIA
VENTRAL NERVE CORD

FIGURE 17.11
EARTHWORM: LATERAL DISSECTION

development in the seminal vesicles. They are passed out through funnel-shaped mouths of the **vas deferens** or **sperm ducts**, which is also well hidden by the vesicles.

The female reproductive tract is composed of **ovaries, egg sac,** and **oviduct,** all unpleasantly difficult to distinguish. Despite the fact that these worms do not have separate sexes, they cannot fertilize themselves. Copulation must occur. But with both sexes in a single animal, any two worms which meet can copulate (obviously a convenient situation). A sperm transfer then occurs in both directions. After transfer, sperm is stored in **seminal receptacles** which are located in segments 9 and 10.

f.   Cross Section of an Earthworm

In a prepared slide, note the following structures: (Refer to Figure 17.12)

1.   **Body Wall** (beginning with the outermost layer)
     a.   **Cuticle:** thin external chitinous layer
     b.   **Epidermis:** outer cellular layer, which contains epithelial cells
     c.   **Circular Muscle Layer:** located just beneath the epidermis, with the fibers cut longitudinally in the section
     d.   **Longitudinal Muscle Layer:** these fibers are arranged in blocks of feather-like bundles extending toward the center; they are cut transversely
     e.   **Peritoneum:** a thin epithelial lining, separating the body cavity from the body wall.
     f.   **Coelom:** the body cavity

153

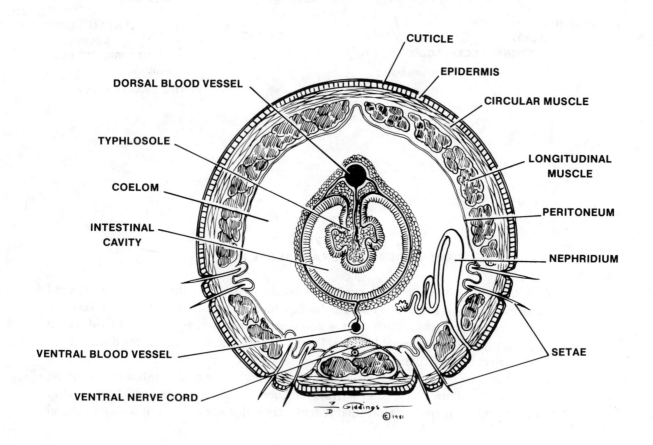

CUTICLE

EPIDERMIS

DORSAL BLOOD VESSEL

CIRCULAR MUSCLE

TYPHLOSOLE

LONGITUDINAL MUSCLE

COELOM

PERITONEUM

INTESTINAL CAVITY

NEPHRIDIUM

VENTRAL BLOOD VESSEL

SETAE

VENTRAL NERVE CORD

**FIGURE 17.12
EARTHWORM (X.S.)**

154

2. **Intestine:**
   a. **Typhlosole:** dorsal invagination of the intestine

3. **Blood Vessels:**
   a. **Dorsal Vessel:** just above the intestine
   b. **Ventral Vessel:** just below the intestine

4. Other Structures:
   a. **Ventral Nerve Cord:** positioned ventrally between the body wall and the ventral blood vessel
   b. **Nephridia:** segmental excretory organs, within the coelom, between the intestine and the body wall; in these sections, only incomplete portions of nephridia can be seen, usually appearing as wavy lines.
   c. **Setae:** 2 pr. ventrally and 2 pr. ventrolaterally projecting from the body wall.

VI. Review Questions

A. List the three major classes of flatworms, and give an example of each.

   1. _Turbellia — Planaria_
   2. _Trematoda — flukes_
   3. _Cestoda — tapeworm_

B. Why do tapeworms not need a digestive system? _____
   _____

C. Identify each of the following terms:

   1. immature proglottid _____

   2. mature proglottid _____

   3. gravid proglottid _____

   4. scolex _____

   5. strobila _____

D. Distinguish between an incomplete and a complete digestive system, and give an example of an animal having each. _____
   _____
   _____

E. Distinguish between a pseudocoelom and a true coelom and give an example of an animal which possesses each. _____
   _____
   _____

155

F. List three examples of parasitic nematodes. _____

_____

G. List four distinguishing characteristics of the Phylum Rotifera. _____

_____

H. What is the function of the typhlosole in an earthworm? _____

_____

I. What is the function of the seminal vesicles in an earthworm? _____

_____

J. How many setae per segment are found in an earthworm? _____

K. What is the function of the earthworm's clitellum? _____

L. Diagram the digestive system of an earthworm, label all of its specialized structures, and give the function of each.

M. Describe four external features by which the ventral surface of an earthworm can be distinguished from the dorsal surface. _____

_____

_____

N. Describe three external features by which one can distinguish the anterior end of an earthworm from the posterior. _____

_____

_____

KINGDOM ANIMALIA:
ARTHROPODA, MOLLUSCA, AND ECHINODERMATA PHYLA

exercise 18

I. Objectives

After the completion of this exercise, the student should be able to do each of the following:

A. Identify the phylum and class of each of the animals in the jars on display.

B. Describe the characteristics contributing to the success of the arthropoda, particularly on land.

C. List the three subphyla of the Phylum Arthropoda, list the classes belonging to each, and cite an example of each class.

D. Dissect a crayfish and identify the parts listed in boldface print in this exercise.

E. Distinguish between a male and a female crayfish.

F. Identify each type of crayfish appendage.

G. List the six classes of Mollusca and give an example of each.

H. Identify the anatomical parts of a starfish listed in boldface in this exercise.

I. Describe the operation of the **water vascular system** in a starfish.

J. List the five classes of Echinodermata and give an example of each.

K. List the features unique to each phylum and class in this exercise.

L. Answer the review questions at the end of this exercise.

II. Phylum Arthropoda

The arthropods are probably the most successful group of animals ever to exist. About four-fifths of all living animal species are arthropods. Their success is attributed to a basic body plan characterized by **segmentation,** a hardened **exoskeleton** with **jointed appendages,** and a high

157

degree of specialization in the brain and central nervous system allowing for a high degree of instinctive behavior. There is great diversity among the animals found in this phylum, though they are usually small in size. They are found in almost every conceivable environment.

The members of several of the classes of this phylum, such as the insects, centipedes, millipedes, and arachnids, are primarily terrestrial. They are better adapted to a land environment than any other invertebrates are, largely because of the following characteristics:

1. a **cuticle** which prevents water loss
2. efficient internal respiratory organs
3. **jointed appendages** with a hard, chitinous **exoskeleton**

Furthermore, some (insects) have developed wings, and their ability to fly has made possible their distribution over the earth. In addition, insects display great variation in specialized mouthparts allowing for chewing, biting, piercing, lapping, or sucking. Such variety allows many insects to share the same habitat without having to compete intensely for food.

Few arthropods are very large due to the restrictions of their exoskeletons. Because the exoskeleton surrounds the body, it must be **molted** periodically to allow for growth. Until the new exoskeleton hardens, the animal is helpless.

Below you will find a simplified scheme of classification for the Phylum Arthropoda. Refer to your textbook for a more detailed description of each taxon. Table I lists the general characteristics of the principal classes.

Phylum Arthropoda
Subphylum Trilobita: no living representatives

Subphylum Chelicerata
    Class Merostomata: horseshoe crabs
    Class Arachnida: spiders, mites, ticks, scorpions, and harvestmen (daddy longlegs)

Subphylum Mandibulata
    Class Crustacea: lobsters, crabs, crayfish, and shrimp
    Class Insecta: butterflies, bees, beetles, mosquitos, etc.
    Class Chilopoda: centipedes
    Class Diplopoda: millipedes

Examine the display jars containing arthropod specimens and attempt to determine the class to which each belongs using Table I as a guide to their characteristics.

A.   Crayfish Dissection

    Obtain a specimen of a crayfish, *Cambarus.*

1.   External Anatomy

    Use Figure 18.1 for the external view. Note the **exoskeleton.** Anteriorly it forms the **carapace,** which covers the dorsal and lateral surfaces of the fused head and thorax, the **cephalothorax.** The posterior part of the body, or **abdomen,** is covered by segmentally arranged chitinous plates. These plates are named according to their positions: **tergum** = dorsal plate; **sternum** = ventral plate; and **pleuron** = lateral plate.
    Examine the stalked **compound eyes,** a pair of **antennules,** a pair of **antennae,** the six pairs of mouth appendages, the large **claws** on the **chelipeds,** the **walking legs,** the **swimmerets** on the abdomen, and the broad **uropods** on the last abdominal segment. These, together with the **telson,** form the fan-shaped tail. All appendages are **serially homologous.** In early development and in basic adult structure they are all alike, even though they often differ in detailed form and function. This basic structure may be examined in one of the swimmerets.

Table I.
PHYLUM ARTHROPODA: GENERAL CHARACTERISTICS OF THE PRINCIPAL CLASSES

Characteristic	Crustacea	Insecta	Arachnida	Chilopoda	Diplopoda	Merostomata
body divisions	usually cephalo-thorax and abdomen	head, thorax, abdomen	cephalo-thorax and abdomen	head with body of similar segments	head, short thorax, long abdomen	cephalothorax and abdomen
paired appendages: antennae	2 pairs	1 pair	none	1 pair	1 pair	none
mouthparts (pairs)	mandibles-1 maxillae-2 maxillipeds-3	mandibles-1 maxillae-1 labia-1	chelicerae-1 pedipalps-1	mandibles-1 maxillae-2 maxillipeds	mandibles-1 maxillae-1 maxillipeds	chelicerae-1 pedipalps-1
legs	1 pair per somite or less	3 pairs on thorax	4 pairs on cephalo-thorax	1 pair per segment	2 pairs per segment	4 pairs on cephalothorax
gas exchange	gills or body surface	tracheae	book lungs and/or tracheae	tracheae	tracheae	book gills
principal habitat	salt or freshwater, few on land	mainly terrestrial	mainly terrestrial	all terrestrial	all terrestrial	marine

159

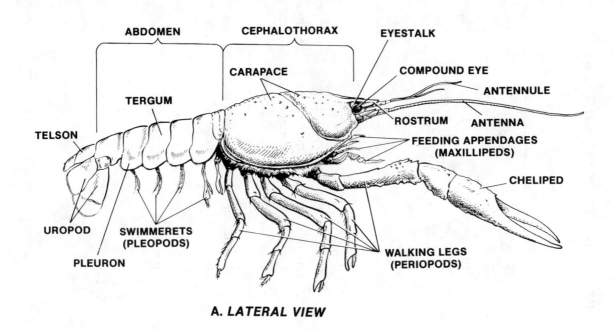

ABDOMEN   CEPHALOTHORAX   EYESTALK

CARAPACE   COMPOUND EYE

TERGUM   ANTENNULE

ROSTRUM   ANTENNA

TELSON   FEEDING APPENDAGES (MAXILLIPEDS)

CHELIPED

UROPOD

SWIMMERETS (PLEOPODS)

PLEURON   WALKING LEGS (PERIOPODS)

A. *LATERAL VIEW*

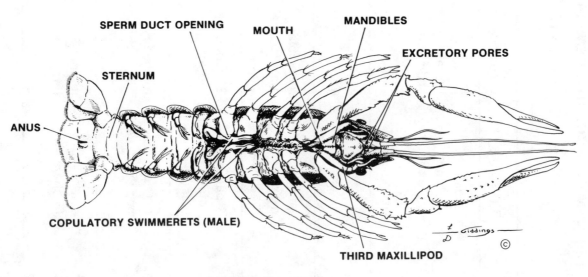

SPERM DUCT OPENING   MOUTH   MANDIBLES

EXCRETORY PORES

STERNUM

ANUS

COPULATORY SWIMMERETS (MALE)

THIRD MAXILLIPOD

B. *VENTRAL VIEW*

FIGURE 18.1
***CAMBARUS* (CRAYFISH): EXTERNAL FEATURES**

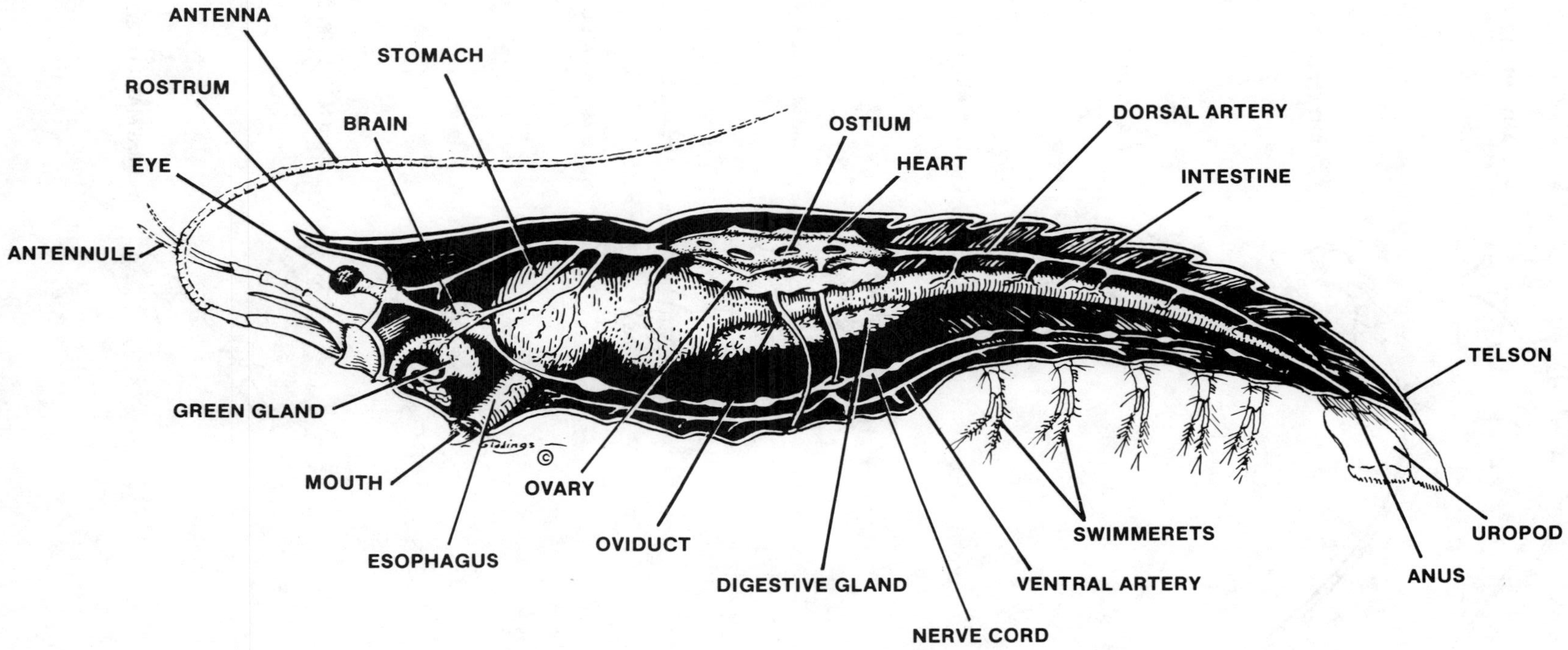

FIGURE 18.2
LONGITUDINAL SECTION OF FEMALE CRAYFISH

ANTENNA
ROSTRUM
EYE
ANTENNULE

STOMACH
BRAIN

GREEN GLAND
MOUTH
ESOPHAGUS

OVARY
OVIDUCT

DIGESTIVE GLAND

NERVE CORD

VENTRAL ARTERY

SWIMMERETS

ANUS

UROPOD

TELSON

DORSAL ARTERY

INTESTINE

OSTIUM
HEART

161

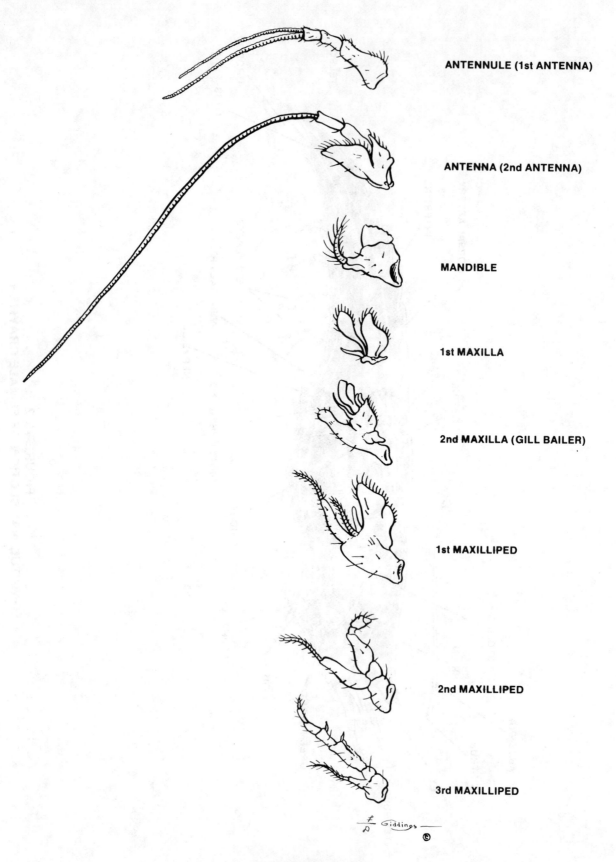

ANTENNULE (1st ANTENNA)

ANTENNA (2nd ANTENNA)

MANDIBLE

1st MAXILLA

2nd MAXILLA (GILL BAILER)

1st MAXILLIPED

2nd MAXILLIPED

3rd MAXILLIPED

FIGURE 18.3
CRAYFISH: HEAD APPENDAGES

162

Remove the appendages on the left side of the crayfish one by one, starting at the posterior end and making sure that the whole appendage is taken off right at the base. Keep track of both the numbers and types of appendages removed, and lay them out in sequence on a sheet of paper. From the abdominal segments you should obtain, posterior to anterior, one telson, one uropod, and five swimmerets. In the female, the anteriormost swimmeret is small or absent; in the male, it functions in sperm transfer and is larger and anteriorly directed. To obtain the thoracic appendages, remove the left side of the carapace with scissors. This will expose a gill chamber and the **gills,** which are attached to all of the thoracic appendages except the first. Remove these thoracic appendages with their gills attached. There will be four similar walking legs and one large cheliped. Next remove the mouthparts. From posterior to anterior there are three **maxillipeds,** two **maxillae,** and one **mandible.** The maxillipeds are more obviously leg-like than the other mouthparts. They are used in sensory functions and in handling and tearing pieces of food. The maxillae are very thin and lie closely pressed to the hard, clublike mandible (jaw). Continue forward and remove in order the second antenna and the first antenna. The first antenna is referred to as an antennule also. Refer to Figure 18.3.

Note the opening at the base of the second antenna. This is the external opening of the **green gland,** the excretory organ of the crustaceans.

2. Internal Anatomy

Refer to Figure 18.2 in order to locate the internal organs. Carefully loosen the remainder of the carapace and the dorsal skeleton of the abdomen from the underlying membranous epidermis. Remove the exoskeleton and cut the epidermis to expose the internal organs. If muscles are in the way, do not tear them, but cut them with scissors. Cover the animal with water, and study the following:

The small **heart,** showing several openings, or **ostia,** is embedded in the pericardial cavity in the middorsal region. Anterior to the heart is the **stomach,** in the head region, a large sac containing a grinding structure, the **gastric mill,** in its wall. Follow the stomach posteriorly and trace the **intestine** to the **anus,** located ventrally in the last abdominal segment. Anteriorly, the stomach leads to a short **esophagus** which passes ventrally to the mouth. Locate the mouth and probe through it to the stomach. To each side of the stomach are the large yellowish **digestive glands.** Behind them, to each side of the heart, are the **gonads.** The **testes** are difficult to distinguish from the digestive glands, but the **ovaries** are coarser in texture and darker in color (almost orange). Ducts from the reproductive organs lead to the exterior openings on the basal segments of the third pair of walking legs in the female, and of the fifth pair of walking legs in the male.

Starting in the abdomen and working forward, carefully remove muscles and other structures to expose the **ventral nerve cords** for their entire length. Note the segmental thickenings of ganglional tissue. Try to identify the ring of nerve tissue encircling the esophagus and leading to the **brain ganglia,** which are the anterior portions of the ring. Also, locate the pad-like **green glands** (not green in preserved specimens) which lie beside the esophagus.

III. Phylum Mollusca

The six classes of molluscs, containing an estimated 110,000 species, are represented by a tremendous variety of body forms, all derived from the same essential body organization. Molluscs are known as the "soft bodied animals."

Features unique to this phylum include:

1. A fleshy epithelial **mantle** that may secrete a calcareous shell.

2. A muscular ventral **foot.**

3. A dorsal **visceral body mass.**

Some additional characteristics of the phylum are:

1. An unsegmented body.

2. An open circulatory system.

3. Respiration by a single or many **ctenidia** (gills), by the mantle, or by the epidermis.

4. An extremely-reduced coelom.

5. Variety in feeding behavior and locomotion. The chiton, for example, is designed for algal grazing and adherence to wave-beaten rocks, the clam for filtering fine food material, the snail for gliding and protection, and the octopus for speed and predation.

See Figure 18.4 for examples of each of the molluscan classes, and examine the specimens on display in jars. Try to determine the class to which each belongs.

A. Class Amphineura

These are the algal-grazing chitons occurring mainly in marine intertidal areas. Some preserved specimens are on display. Note the **foot, gills,** and the 8 linked calcareous plates that form a protective shell.

B. Class Monoplacophora

These are very primitive, somewhat segmented molluscs, known from 300 million year old fossils and from living specimens discovered in 1952 off the coast of Central America.

C. Class Bivalvia

Bivalves, such as clams, oysters, mussels, and scallops comprise this group, all of which have shells comprised of two halves called **valves.** These dorsally hinged valves are tightly closed by well-developed adductor muscles. In general, bivalves are water-filtering organisms enclosed in shells.

Identify the **mantle, foot,** and **gills** of the clam or oyster on demonstration.

D. Class Gastropoda

This class includes snails, whelks, limpets, slugs, and nudibranchs. Except for the slugs and nudibranchs, most gastropods possess a single spiral shell. These molluscs have well developed heads with tentacles and a rasping **radula** — a structure which enables them to chew up vegetation.

E. Class Cephalopoda

The cephalopods are highly modified for motility and active predation. These, such as squid and octopus, possess a large head and eyes, a highly developed brain and central nervous system, 8, 10, or more arms equipped with rows of sucking discs, a mouth with horny beak and radula, and a large **siphon** for controlled, rapid movement. The arms, or tentacles, correspond to the foot in other molluscs.

164

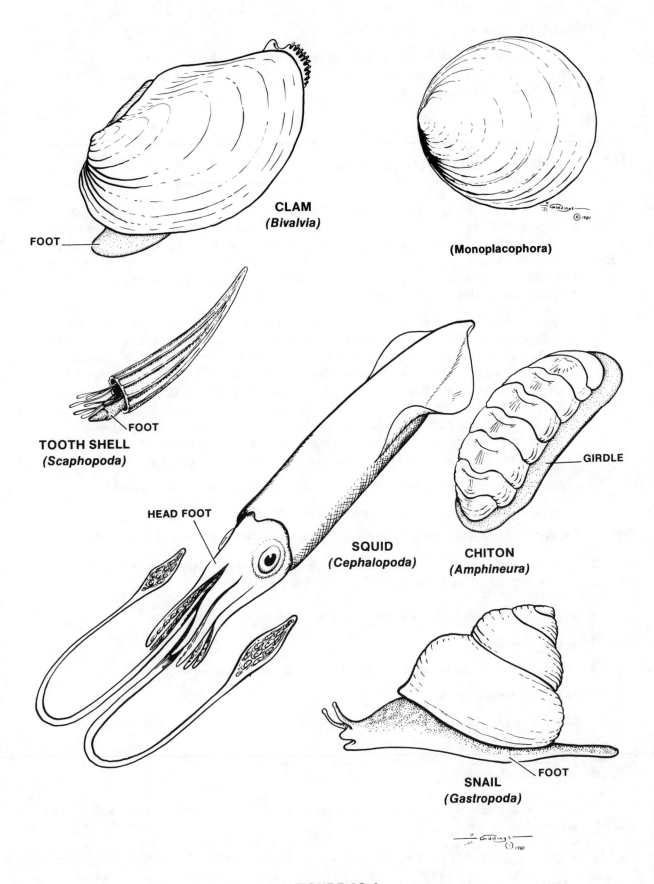

CLAM
(Bivalvia)

FOOT

(Monoplacophora)

TOOTH SHELL
(Scaphopoda)

FOOT

HEAD FOOT

SQUID
(Cephalopoda)

GIRDLE

CHITON
(Amphineura)

SNAIL
(Gastropoda)

FOOT

FIGURE 18.4
REPRESENTATIVE MOLLUSCS

165

F. Class Scaphopoda

   The elephant-tusk shells are a small group adapted to life in mud or sand in marine waters. They are very seldom seen alive. Tusk, or tooth shells are identifiable by their tubular shells which are open at both ends.

IV. Phylum Echinodermata

   Included in this phylum are the spiny skinned animals, such as the sea lilies, starfish, brittle stars, sea urchins, and sea cucumbers. All of the species are marine. Adult forms are sessile or slowly creeping forms which are radially symmetrical around an oral-aboral axis. The larvae are free-swimming, bilaterally symmetrical forms. Echinoderms appear to have evolved from an ancestral line having bilateral symmetry such as the flatworm or some similar organism.
   Features unique to this phylum include:

1. A **water vascular system.** This is a system of internal tubes communicating with the exterior by way of a sieve plate or madreporite, which regulates the amount of water in the system, and ending in a paired series of tube feet running the length of each ray, or along each section of the fused **endoskeleton,** or **test,** in sea urchins. The tube feet are extended by the contraction of muscular bulbs, **ampullae,** at their inner ends which force water into the tube feet, making them turgid. When the tube feet are brought in contact with a surface, the ampullae relax, permitting the echinoderm to adhere strongly to a surface without further use of energy. This enables them to withstand the crashing surf in the intertidal zone as well as to open the shells of bivalve mollusks for food.

2. Minute respiratory structures, skin gills, **dermal papillae,** or **dermal branchia.**

3. A calcareous endoskeleton of movable or fixed plates or ossicles.

4. Numerous hard spines that arise from the internal skeleton.

5. Many minute pincers, or **pedicellaria** which act to keep the body surface free of debris, aid in capturing food, and protecting the skin gills.

   Some additional characteristics of the phylum are:

1. Ciliated organs.

2. Nervous system consisting of a circumoral ring and radial nerves to the arms.

3. Lack of cephalization.

4. Complete digestive tract.

5. Lack of segmentation.

6. Open circulatory system.

   Living representatives are divided into 5 classes. Refer to Figure 18.5. Also examine the many specimens on display.

A. Class Echinoidea

   These spiny sea urchins and sand dollars are herbivorous echinoderms constructed as though their arms were folded back into a ball and fixed into a calcareous skeleton or test,

166

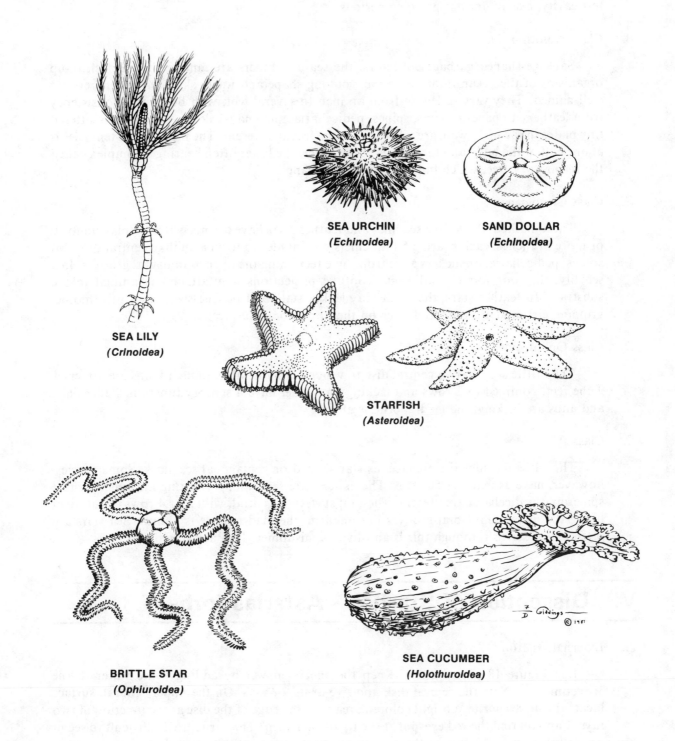

SEA LILY
(Crinoidea)

SEA URCHIN
(Echinoidea)

SAND DOLLAR
(Echinoidea)

STARFISH
(Asteroidea)

BRITTLE STAR
(Ophiuroidea)

SEA CUCUMBER
(Holothuroidea)

FIGURE 18.5
REPRESENTATIVE ECHINODERMS

then covered with long, sharp, movable spines and 3-jawed pedicellaria. The test is globular in sea urchins, and disc or heart-shaped in sand dollars. Tube feet are long, slender and equipped with suckers; mouth and anus are central or lateral. The large gut fills much of the test cavity, except during spawning periods.

B. Class Holothuroidea

Sausage-shaped garbage collectors, the sea cucumbers are among the chief clean-up organisms of the ocean floor. With their oblong shape and warty skin, sea cucumbers are well-named. They vary in length from an inch to several feet, with body wall consistency from leathery to papery. Arms, spines, pedicellaria, and endoskeleton, except for scattered tiny plates in the body wall, are all absent. Tube feet are present. The mouth with tentacles is at one end of the body and the anus is at the other, the latter often bearing a complex called the **respiratory tree,** which functions in gas exchange.

C. Class Crinoidea

These stalked, flowerlike sea lilies and feather stars have 5 arms which display up to 10 or more branches, each bearing 5 branchlets or pinnules to form a cuplike central disc. No spines, pedicellaria, or suckers arise from tube feet lining the open ambulacral grooves. In a sea lily, the long jointed stalk with rootlike projections may attach the animal to the substrate. In feather stars, the adult may lack a stalk and be free-swimming with motile, gripping cirri and a mouth and anus on the upper, **oral,** surface.

D. Class Ophiuroidea

The brittle stars have a central disc to which highly flexible, jointed limbs are attached. Tube feet, confined to 2 rows and lacking ampullae, have a sensory function. Pedicellaria and anus are lacking; the madreporite is aboral.

E. Class Asteroidea

This class includes the predaceous star-shaped or pentagonal sea stars. Some species, however, have as many as 50 arms. The ossicles are separate, permitting movement; short spines and pedicellaria are present. The oral surface is ventral, with 2 or 4 rows of tube feet lining the open ambulacral grooves in each arm; the madreporite is **aboral.** Sea stars are often called starfish, though this is an obvious misnomer. Fish are vertebrates.

V.   Dissection of Starfish - Asterias forbesii

A. External Anatomy

Use Figure 18.6 as a guide. Keep the specimen wet by adding some water to the dissecting pan. Note the central disk and five arms, or **rays.** On the upper, **aboral,** surface locate the **madreporite,** a bright colored area near the edge of the disc at the junction of two rays. Can you find the red **eyespot** at the tip of each arm? This structure is difficult to see in preserved specimens. Observe the hard **spines** scattered over the surface. Located among the spines are the **dermal branchiae** and pincerlike **pedicellaria.**

On the oral or ventral surface, locate the **mouth,** surrounded by large spines. Is there any material protruding from the mouth? If so, what might it be? Running along the middle of each ray is an **ambulacral groove,** with rows of **tube feet.** Identify the suction cups at the ends of the tube feet.

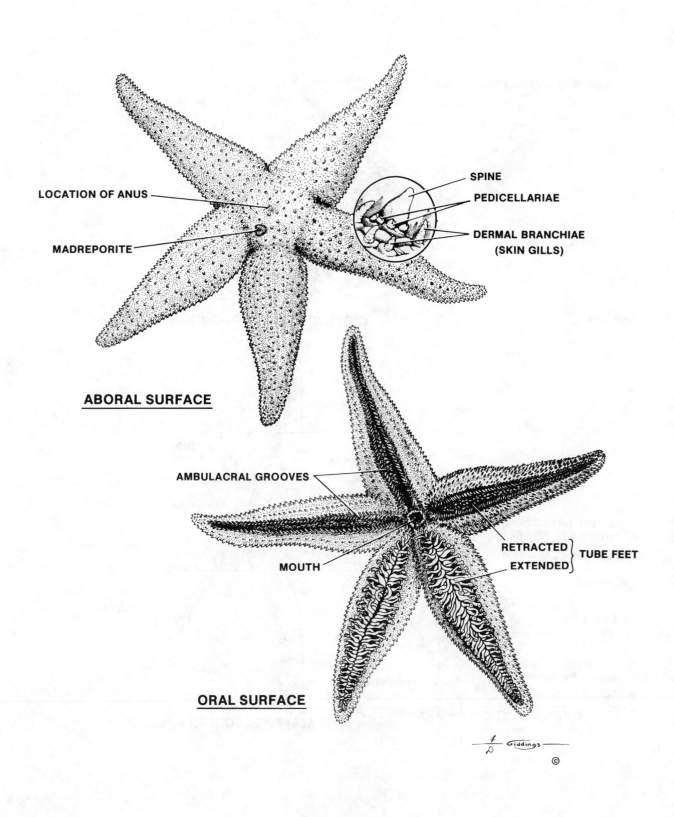

LOCATION OF ANUS

MADREPORITE

SPINE

PEDICELLARIAE

DERMAL BRANCHIAE
(SKIN GILLS)

ABORAL SURFACE

AMBULACRAL GROOVES

MOUTH

RETRACTED }
EXTENDED } TUBE FEET

ORAL SURFACE

Giddings
©

FIGURE 18.6
ASTERIAS **(STARFISH): EXTERNAL FEATURES**

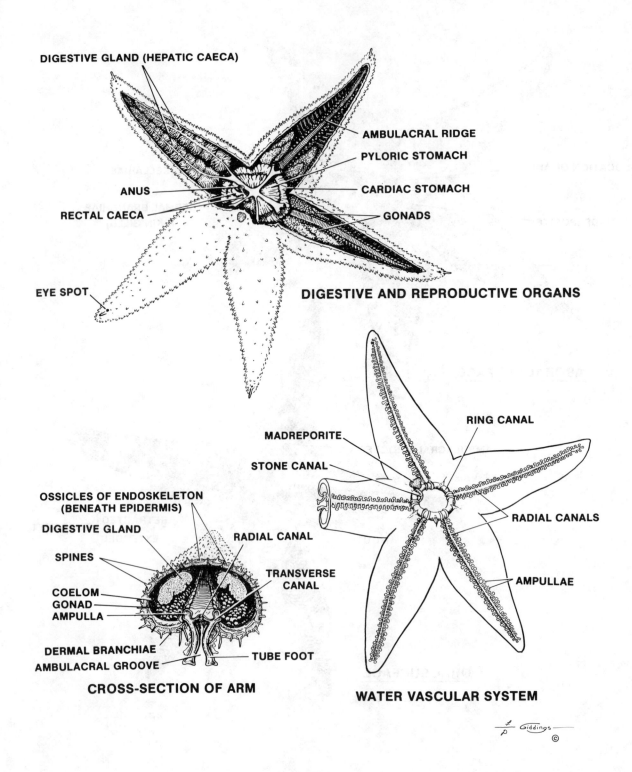

DIGESTIVE GLAND (HEPATIC CAECA)

AMBULACRAL RIDGE

PYLORIC STOMACH

ANUS

CARDIAC STOMACH

RECTAL CAECA

GONADS

EYE SPOT

DIGESTIVE AND REPRODUCTIVE ORGANS

MADREPORITE

RING CANAL

STONE CANAL

RADIAL CANALS

OSSICLES OF ENDOSKELETON
(BENEATH EPIDERMIS)

DIGESTIVE GLAND

RADIAL CANAL

SPINES

TRANSVERSE
CANAL

AMPULLAE

COELOM
GONAD
AMPULLA

DERMAL BRANCHIAE

AMBULACRAL GROOVE

TUBE FOOT

CROSS-SECTION OF ARM

WATER VASCULAR SYSTEM

Giddings ©

FIGURE 18.7
ASTERIAS **(STARFISH): INTERNAL FEATURES**

B.   Internal Anatomy

Cut off about one-half inch of the tip of the ray farthest from the madreporite and cut through the aboral wall of the ray along each side toward the central disc. Repeat this on an adjacent arm. Continue cutting around the central disc such that a circular area is removed from the disc, leaving only the madreporite and anus. Work carefully so that the delicate organs beneath are not macerated. Then, beginning near the tip of the ray, **carefully** remove the aboral skeleton in small sections, lifting and freeing it from the underlying tissue before actually cutting off the skeleton. Carefully examine Figure 18.7 showing a partial dissection.

The **mouth** leads into a short esophagus which is connected to a much-folded, sac-like **cardiac stomach** which is the portion that can stick out through the mouth of the starfish and start digesting the contents of an oyster. Above the cardiac stomach is another portion of the stomach known as the **pyloric stomach.** Connected to this are five **digestive glands,** each of which is located in a ray. Dorsal to the pyloric stomach are several **rectal caeca** plus a short intestine attaching the stomach to the **anus.** The rectal caeca and anus are difficult to find.

A pair of **gonads** is located in each ray. They are located underneath the digestive glands and are usually a deep brownish-red color and smaller in size than the digestive glands.

After finding the gonads and digestive glands, remove them from one of the rays. Notice the skeletonous **ambulacral ridge** running down the center of the ray. Along each side of the ridge are tiny reddish bulblike structures — these are the **ampulla.** Also, near where the arms are attached to the central disc, find the **two bands of muscle** per arm, responsible for movement of the arms.

To see the **water vascular system,** use Figure 18.7. Remove the stomach and try to find the **stone canal** leading orally from the madreporite to the **ring canal,** a hard ring around the mouth, from which branch 5 **radial canals** (one into each arm). Split the ambulacral ridge lengthwise to find the radial canal. It is connected to the **ampulla** and **tube feet** via short lateral canals.

VI.   Review Questions

A.   List the functions of the following appendages of arthropods:

1.   cheliped _____

2.   antenna _____

3.   wing _____

4.   maxilliped _____

5.   maxilla _____

6.   mandible _____

7.   swimmeret _____

8.   uropod _____

B.   What is the function of the green glands of the crayfish? _____

_____

C. List the adaptations of arthropods for success in a terrestrial environment. _____

_____

_____

D. What is the literal meaning of the word "arthropoda"? _____

E. In what respects do the Chelicerata and the Mandibulata differ? _____

_____

_____

F. What is the function of the water vascular system found in starfish? _____

_____

G. Identify each of the following structures with respect to function and the phylum with which it is associated:

radula _____

ampulla _____

madreporite _____

siphon _____

compound eye _____

respiratory tree _____

KINGDOM ANIMALIA:
HEMICHORDATA AND CHORDATA PHYLA

exercise 19

I. Objectives

After the completion of this exercise, the student should be able to do each of the following:

A. Explain the evolutionary relationships thought to exist between the echinoderms, hemi-chordates, and chordates.

B. Identify the **proboscis, collar, trunk,** and **pharyngeal gill slits** in a prepared slide of *Balanoglossus*.

C. Identify the phylum and class to which each of the animals in the jars on display belong.

D. List the features unique to each phylum, subphylum, and class discussed in this exercise.

E. List at least four adaptations exhibited by reptiles for life on land.

F. List four specialized adaptations for flight exhibited by birds.

G. Identify the chordate features which are lost by tunicates in the transition from larva to adult.

H. Identify the **buccal cirri, dorsal nerve cord, notochord, gill bars, myotomes, atrium,** and **atriopore** in a prepared slide of an amphioxus.

I. Identify any of the parts given in boldface in the dissection of the frog and give the function of each.

J. Answer the review questions at the end of this exercise.

II. Phylum Hemichordata

Present-day hemichordates are wormlike types known as **enteropneusts,** or **acorn worms.** They can be found burrowing in the sand and mud of tidal flats and shallow coastal waters. *Balanoglossus* and *Dolichoglossus* are common genera along the North American coasts.

Examine the plastomounts, slides, and preserved specimens in the jars on display. The body of the worm is divided into an anterior muscular **proboscis,** a short **collar,** and a posterior elongate **trunk. Pharyngeal gill slits** can be seen in the anterior region of the trunk. This is a characteristic hemichordates share with the chordates, along with the possession of a **dorsal nerve cord.** However, in hemichordates, the nerve cord is very short and is restricted to the collar region. A cartilaginous rod is also found in both hemichordates and chordates, but in the former it does not underlie the nerve cord as it does in the latter. For this reason, it is given the name **stomochord** in the hemichordates, while it is called a **notochord** in the chordates. Refer to Figure 19.1.

Acorn worms begin life as bilaterally symmetrical, ciliated larvae very much resembling those of the echinoderms. And their coeloms are divided, again a characteristic more resembling echinoderms than chordates. It is for all of the above reasons that the hemichordates, once classified as a subphylum of the Phylum Chordata, are now placed in a separate phylum by most taxonomists.

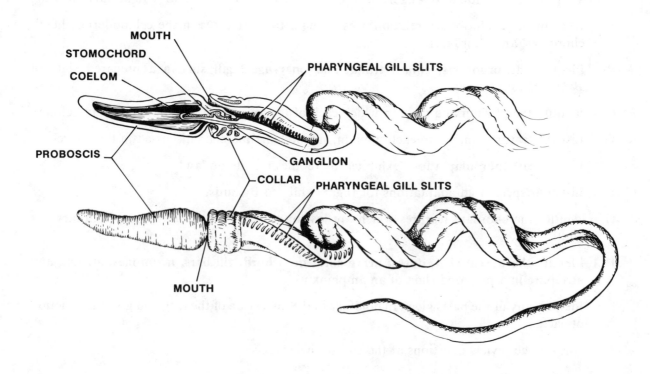

FIGURE 19.1
***BALANOGLOSSUS* (EXTERNAL VIEW AND SAGITTAL SECTION)**

174

III. Phylum Chordata

This is the most advanced phylum, including a tremendous variety of organisms ranging from wormlike animals up to and including man. Despite the variation, all chordates are characterized by three unique features at some stage in their life histories. Sometimes these features may appear only in the embryonic life and disappear before the organism matures. These three features are as follows:

1. A **notochord,** a flexible supportive rod extending the length of the body dorsal to the digestive tract and furnishing skeletal support. The notochord is replaced by a backbone in the subphylum Vertebrata.

2. A **hollow nerve tube,** located **dorsal** to the notochord.

3. **Pharyngeal gill slits,** which function in respiration in the lower chordates but appear only in the developing embryo of higher chordates.

The three chordate subphyla are listed as follows:

A. Subphylum Tunicata or Urochordata

These are the tunicates or sea squirts. They are highly modified, sessile, filter-feeding animals with motile larvae. The larva shows all three chordate characteristics, but in the transition to adulthood, all of the notochord and most of the nerve cord are reabsorbed with the tail. An unusual feature of the adult is that the tunic surrounding the animal contains **cellulose,** a substance usually associated with plants. As in the acorn worms, the gills serve in feeding as well as in gas exchange. Food particles carried in by water entering the **incurrent siphon** are trapped in mucus on the **gill bars** as the water goes through the **gill slits** into the **atrium.** The water then leaves the atrium via the **excurrent siphon.** Refer to Figure 19.2 as you examine plastomounts and specimens of *Molgula* and its relatives in the jars on display.

B. Subphylum Cephalochordata

The lancets, or amphioxus, are small, fishlike, mud or sand-dwelling filter feeders, important from a theoretical standpoint. They represent an evolutionary stage in which chordate characteristics are well developed, but vertebrate characteristics are absent. Amphioxus is a chordate of great interest, a prototype of how an ancestral vertebrate might have looked. Except for annelidlike ciliated nephridia, the lancet's organ systems are similar to those of vertebrates or other simple chordates.

Study the internal anatomy of **amphioxus** by examining a stained slide specimen, while referring to Figure 19.3. Also examine the specimens in plastomounts and display jars.

C. Subphylum Vertebrata or Craniata

This is the largest group of chordates and consists of animals in which the notochord is replaced by a vertebral column as the primary supporting structure. They have a brain consisting of five hollow vesicles, a jointed endoskeleton with a cranium to enclose the brain, a backbone of segmented vertebrae, and the concentration of sensory organs in the head region, a phenomenon known as **cephalization.**

The subphylum vertebrata includes the following classes:

1. AGNATHA

These jawless fishes are the most primitive vertebrates and lack limbs and jaws. The first records of vertebrates from early fossil deposits show they were abundant members of the sea-bottom fauna about 400 or more million years ago. Fossil agnatha

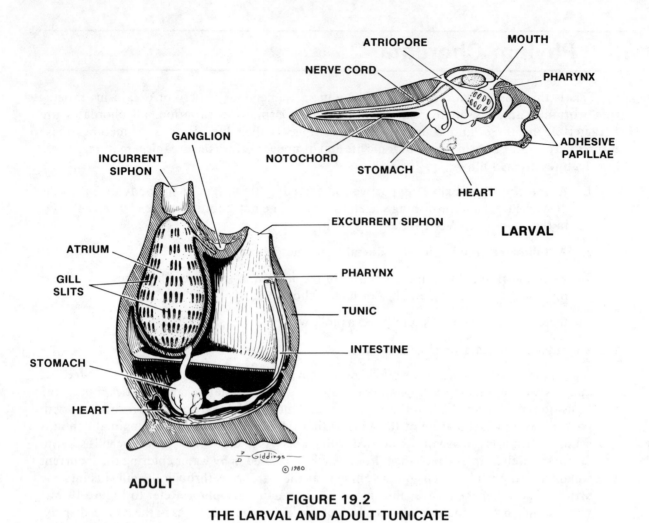

GANGLION

INCURRENT
SIPHON

ATRIOPORE   MOUTH

NERVE CORD

PHARYNX

NOTOCHORD

ADHESIVE
PAPILLAE

STOMACH

HEART

LARVAL

EXCURRENT SIPHON

ATRIUM

PHARYNX

GILL
SLITS

TUNIC

INTESTINE

STOMACH

HEART

ADULT

FIGURE 19.2
THE LARVAL AND ADULT TUNICATE

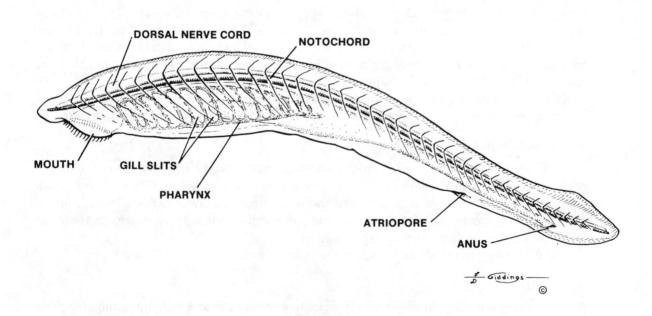

DORSAL NERVE CORD   **NOTOCHORD**

MOUTH

GILL SLITS

PHARYNX

ATRIOPORE

ANUS

FIGURE 19.3
AMPHIOXUS

(Ostracodermi) were characterized by heavy bony armor and probably a mud-sucking, filtering mode of feeding. Their highly specialized descendants, the lampreys and hagfishes *(Cyclostomata),* are still in existence. Survival of modern remnants of this ancient class is probably related to their specially adapted rasping and bloodsucking mouthparts, which enable them to feed on other fish. This parasitic way of life frees the lamprey from competition with later-evolved forms. Larval lampreys, or **ammocoetes larvae,** are good illustrations of the primitive vertebrate form, being strikingly similar to amphioxus.

Besides lacking jaws as adults, agnathans lack paired appendages and have a cartilaginous skeleton, and a single unpaired nostril. Examine the specimens on display.

2. PLACODERMI

These armored fish are now entirely extinct. The class is marked by development of a primitive type of jaw suspension that enabled the placoderms to replace the more primitive jawless fishes, although they themselves died out in the Permian period approximately 240 million years ago.

3. CHONDRICHTHYES

This class of cartilaginous fish includes the sharks, rays, skates, and chimaeras. It is doubtful whether these fish preceded true bony fish in evolution, as is often assumed. Perhaps the two groups evolved in parallel fashion from placoderm ancestors.

The all-cartilaginous skeleton appears to be a comparatively recent adaptation (though still several hundred million years old). Another important adaptation in the group is the capacity to retain **urea** in the blood and body tissues, thus raising the internal osmotic pressure to a point nearly equal to that of the salt-water environment.

Chondrichthyes can be distinguished from agnathans by their smaller number of gill openings (5 pair), their paired nostrils, and their paired **pectoral** and **pelvic fins.** Another feature worthy of note is that their skins are covered by tiny **placoid scales,** structures which are tooth-like in their interal anatomy and which are thought to be the evolutionary precursors of the teeth, scales, hair and feathers characteristic of the following classes. Examine the specimens of chondrichthyes on display.

4. OSTEICHTHYES

The enormously varied and numerous bony fish represent some 75 percent of described vertebrate species, including nearly all freshwater fish and the preponderance of marine species. The most distinctive external feature of this group is the possession of a gill cover, or **operculum,** operated by muscles and allowing the aeration of the gills in a bellows-like fashion while the fish is sitting still. The scales of bony fish are more elaborate and larger than those of the chondrichthyes.

One subclass of bony fish of special importance to students of evolution is the Choanichthyes (lobe-finned fishes), fish with nostrils connected to the mouth cavity and with paired fleshy or limb-like fins. These are presumed precursors of the amphibians and very ancient groups which have changed little since. Familiar lobe-finned fishes include the lungfish and the remarkable "living fossil", *Latimeria,* discovered in the fifties off the coasts of South Africa and Madagascar. Until the discovery, this fish was presumed to have been extinct for over 75 million years. Examine the specimens of bony fish on display.

5. AMPHIBIA

Salamanders, frogs, toads, and caecilians (or apoda-tropical, limbless amphibians) were the first tetrapoda (possessing four limbs), or land-dwelling, limbed vertebrates. Amphi-, meaning "both", implies life both in water and on land.

Amphibians require abundant moisture, as their soft skin is not protected against water loss. Nearly all amphibians depend on water for egg laying and early development, though some frogs breed in damp moss instead of water. Certain toads are remarkably resistant to dessication; other amphibians such as the Mexican axolotl and a few frog species never leave the water. In general, we think of amphibians as transitional creatures, perfectly adapted neither to water nor to land, but somewhat adapted to both. Organisms like these could have been the ancestors of purely terrestrial animals.

Some characteristics of amphibians include:

a. Skeleton largely bony.

b. Ribs, if present, are not attached to the sternum.

c. Three chambered heart consisting of two **atria** and one **ventricle.**

d. Gas exchange by gills, lungs, skin, or lining of the oral cavity.

e. 10 pairs of cranial nerves.

f. **Poikilothermy** (dependence of body temperature on environment)

g. Fertilization external or internal. Mostly oviparous — egg layers. Eggs without shells.

h. **Holoblastic cleavage,** in other words, the whole zygote divides.

In a later section we will study the anatomy of the familiar leopard frog, *Rana pipiens,* a typical member of this class.

6. REPTILIA

Snakes, lizards, turtles, crocodiles, and alligators, as well as numerous extinct forms (many different groups are linked by the common name "dinosaur") are included in this class. Reptiles are thought to have evolved from amphibian ancestors called labyrinthodonts. These organisms represent the completed transition from water to land. They are successful on land for a number of reasons:

a. The dry scaly skin helps to conserve moisture.

b. Internal fertilization to improve egg survival.

c. They form **amniotic eggs,** in which the embryo is enclosed in a fluid-filled sac. These are covered by leathery shells.

d. They have lungs for gas exchange in an air medium.

e. A free larval stage is lacking.

f. Nitrogenous wastes are excreted primarily as **uric acid,** an insoluble solid which can be stored within the egg shell.

g. More efficient limbs and superior muscular coordination.

h. Poikilothermy

i. 12 pairs of cranial nerves.

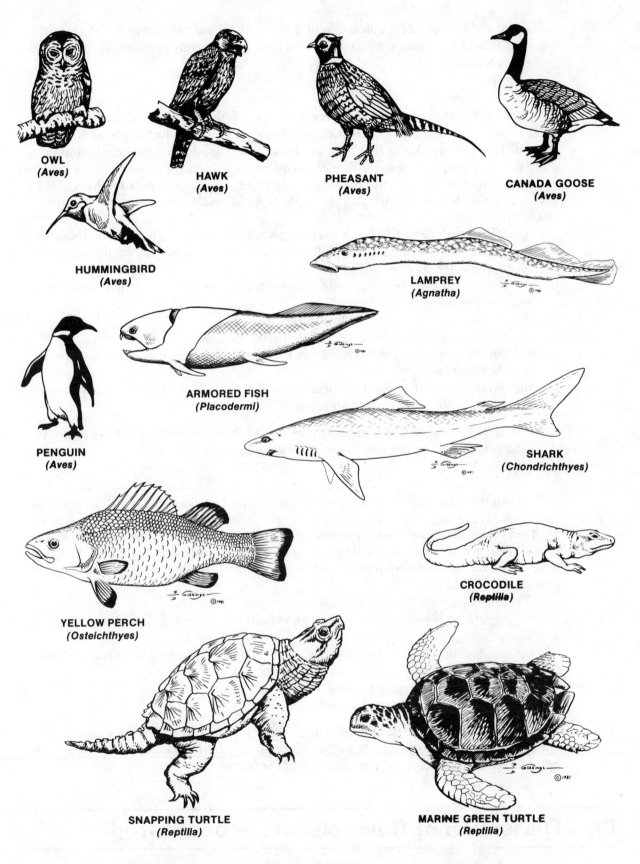

FIGURE 19.4
REPRESENTATIVES OF THE CLASSES OF SUBPHYLUM VERTEBRATA
(AMPHIBIANS AND MAMMALS NOT INCLUDED).

j.   Three chambered heart consisting of 2 atria, and a partially chambered ventricle, except in crocodiles and alligators which have a completely partitioned ventricle or two ventricles.

7.   AVES

Birds have retained many of the features of the reptiles, from which they evolved, such as the formation of an amniotic egg, lungs, and internal fertilization. However, birds differ from reptiles in that they are **homeothermic,** that is, they can maintain a constant body temperature in the face of environmental fluctuation and thus can remain active throughout the year. A covering of modified scales, known as **feathers** aids in preventing heat loss by creating a lightweight layer of insulation. Birds also have four-chambered hearts.

Certain of the adaptations of birds are specializations for flight. Among them are:

a.   the development of **pneumatic** (hollow) bones connected by way of air sacs to the lungs makes the bird much lighter.
b.   the development of a one-way system for ventilation of the lungs which makes more rapid gas exchange possible.
c.   the alteration of the forelimbs into **wings.**
d.   the development of powerful **pectoral muscles** for operation of the wings.
e.   the development of a **high-keeled breastbone** to increase the area available for muscle attachment.
f.   the development of **uncinate processes** on the ribs which strengthen the ribcage and provide for more muscle attachment.

Examine the pigeon skeleton on display for the skeletal modifications listed above. Also examine the preserved specimens of birds in the jars on display.

8.   MAMMALIA

Mammals evolved from a different subclass of reptiles than that which gave rise to the birds. Like birds, mammals are homeothermic. Gas exchange involves lungs assisted by a muscular **diaphragm,** and fertilization is internal. Some mammals, such as the duck-billed platypus and the spiny anteater lay eggs with shells, that is, they are **oviparous.** But most are **viviparous,** or bring forth live young. Mammals also have four-chambered hearts.

The distinguishing features of this class include:

a.   the possession of modified scales known as **hair,** which can function as an insulating layer.
b.   the production of milk by modified sweat glands known as **mammary glands,** providing a source of food for the young.
c.   the possession of a highly developed nervous system.
d.   the possession of a muscular diaphragm.

Examine the specimens of mammals in the jars on display. Then examine all jars containing vertebrate specimens and attempt to determine the class to which each belongs on the basis of its observable external features.

IV.   Dissection of Rana pipiens, leopard frog

The leopard frog is the species most frequently used for laboratory work. The bullfrog, *Rana catesbeiana,* will have been dissected for display in the lab. Use the bullfrog for comparison with the leopard frog.

A. External Anatomy

Notice that the body of the frog can be divided into a **head** and a **trunk**. A distinct neck is absent (a characteristic retained from fish, since the presence of a neck would allow for independent movement of head and trunk — an unnecessary movement when swimming.).

A large **mouth,** a pair of **external nares** (nostrils), and **eyes** are located on the head. Associated with the eye are an upper **eyelid** that is just a fold of skin and a lower eyelid, a transparent membrane, that can be extended over the surface of the eyeball. On the head, behind the eye, is a disc-shaped structure, the **tympanic membrane;** this functions as the eardrum.

Examine the front leg of the frog, noting the **upper arm,** the **elbow,** the **forearm,** the **wrist joint,** and the **hand** with 4 **fingers.** The finger closest to the body is thicker than the others and is therefore considered to be the thumb. It becomes swollen and deeply pigmented in the male during the breeding season. Looking at the hind leg, note the **thigh, knee, shank, ankle joint,** and the long **foot** with 5 toes. Just dorsal to where the hind legs join at the posterior end of the trunk is the opening common to both the digestive and the urogenital systems. This opening is the **cloacal opening.**

B. Internal Anatomy

Open the mouth, cutting through the angle of the jaw on each side, extending the cut posteriorly to the tympanic membrane. Pull the floor of the mouth ventrally, exposing the **buccopharyngeal cavity.** This cavity consists of the **buccal** (oral) **cavity** in the mouth and surrounded by jaws and the **pharynx,** which is a region between the buccal cavity and the **esophagus,** a tube leading to the stomach. With the help of Figure 19.5 locate and study the following structures: **maxillary teeth** on the margin of the upper jaw; the **vomerine teeth** located on the palate or roof of the buccal cavity (the teeth are used for holding food); the **internal nares,** openings from the nostrils; the **eustachian aperture** (opening) into the eustachian or auditory tube; the **vocal sac apertures** in males; the **glottis,** an opening into the trachea which leads to the lungs; the **esophagus** lying dorsal and posterior to the glottis; the large **tongue** which is attached at its anterior end.

Place the frog ventral side up in a dissecting pan. Fasten the frog to the pan with pins through the tip of the snout and all the limbs. Pick up the skin with forceps and make a longitudinal incision through it, about one-tenth inch to one side of the median line, from the cloacal opening to the lower jaw. Then cut the skin at right angles to the first incision, making these cuts just behind the forelimbs and just in front of the legs. Turn back the two skin flaps and pin them to the pan. Note how loosely the skin is attached to the underlying muscles and how evident the blood vessels are in the underside of the skin.

The dark line of the midventral region of the abdominal muscle layer is the **abdominal vein.** Keep this and other major blood vessels intact as long as possible. Make a longitudinal incision in the body wall on each side of this vein from the **pelvic girdle** to the **pectoral girdle** (the chest area). Make another set of right angle incisions and pin back the muscle flaps. Lift up the pectoral girdle, freeing the structures adhering to it, and cut through the bone.

The **coelom,** or body cavity, is now exposed. It is lined by a shiny layer of epithelial tissue, **peritoneum,** which extends from the middorsal and midventral lines of the body to the organs. The membrane suspending the organs is called the **mesentery.** The coelom allows for the movement of the internal organs; the mesenteries restrict the movement of the organs and also contain the blood vessels and nerves involved with the organs. The frog's coelom is divided into two regions; the large **pleuroperitoneal cavity** in which respiratory, digestive, reproductive and excretory organs are located and the smaller **pericardial cavity** in which the heart is located.

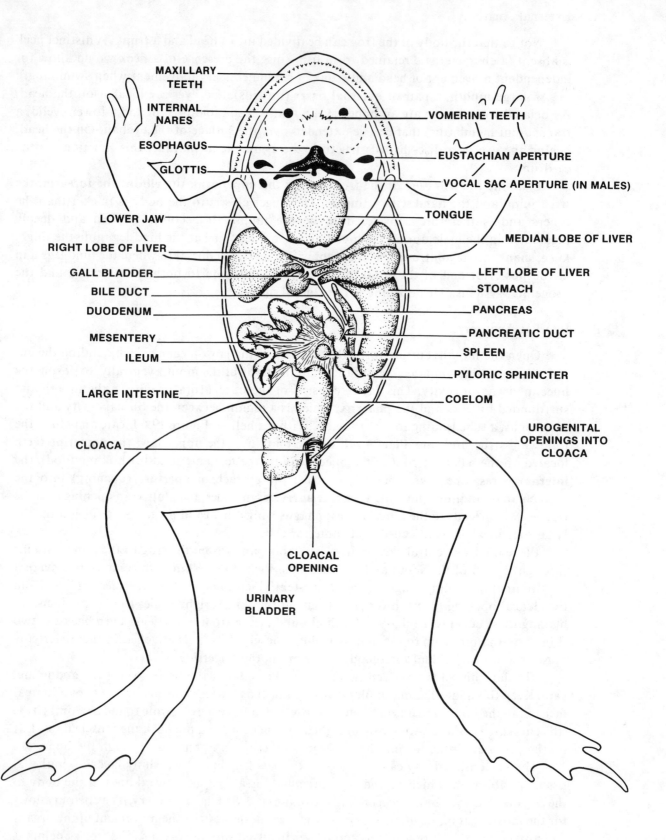

FIGURE 19.5
BUCCAL CAVITY AND DIGESTIVE SYSTEM

1. Digestive System

   Pass a probe down the esophagus and feel with your fingers when the probe enters the **stomach,** a large, sac-like structure in which food is stored and digestion is begun (refer to Figure 19.5). At the posterior end of the stomach is a thick muscular region, the **pyloric sphincter,** which regulates the passage of food from the stomach into the small intestine. Cut the stomach open and note the longitudinal folds. When the pyloric sphincter relaxes, food is allowed to move into the **duodenum,** the first section of the **small intestine.** Cut this tube open and note the folds here also. In addition to secretions from mucus glands in the lining of the small intestine, secretions from the **pancreas** (a whitish tissue lying in the loop between the duodenum and the stomach) enter the duodenum via the **pancreatic duct.** Secretions from the **liver** (a large, three lobed organ in the anterior part of the cavity) and the **gall bladder,** a storage sac for bile, located on the dorsal surface of the liver, enter the duodenum via the **bile duct.** Following the length of the long, thin, coiled small intestine, note that it joins the short fat **large intestine,** which joins with the **cloaca.** (To see this, carefully cut through the bony pelvic girdle and spread the hind legs apart.)

2. Respiratory System

   Find the **glottis** and pass a probe into it until you can feel the probe when it enters the **lungs.** Use Figure 19.5 to help you find the lungs. In preserved frogs, the lungs are contracted to some degree. Enlarge the glottis by making a short longitudinal cut extending anteriorly and posteriorly from it. Spreading it open, you should be able to see a longitudinal fold on each side within the chamber; these are the **vocal folds** used in croaking. The openings to the lungs are beneath these folds.

3. Urogenital System

   The excretory and reproductive organs are closely associated and together they comprise the urogenital system. Before you start to dissect your specimen, refer to Figure 19.6 and Figure 19.7 in order to determine the sex. The excretory structures are similar in the two sexes of the frog.

   A pair of elongated organs, the **kidneys,** are directly attached to the dorsal body wall anterior to the cloaca. They are covered ventrally by the peritoneum. Each kidney is composed of several thousand tubules which filter waste material from the blood. Urine leaving the kidney is transported through the **ureter** to the dorsal surface of the cloaca. The **urinary bladder** is attached to the ventral surface of the cloaca, thereby necessitating the flow of urine across the cloaca before being stored in the bladder. On the ventral surface of the kidney is a light-colored band of tissue, the **adrenal gland.**

   In the male (refer to Figure 19.6), find two small yellowish structures attached by short mesenteries to the ventral surface of the kidneys. These are the **testes,** the site of sperm production. Sperm are transported from the testes via tiny tubules, vasa efferentia, to the kidneys through which they pass before entering the **ureter.** The sperm are carried down to the end of the ureter where they are stored temporarily in an expanded portion called the seminal vesicle. At the time of **amplexus,** the mounting of the female by the male, sperm are released through the male's cloaca at the same time as the female releases the eggs. Fertilization is external. In the leopard frog note the rudimentary **oviducts** extending laterally alongside the kidneys. These are not found in the bullfrog.

   In the female (refer to Figure 19.7), find two multi-lobed organs attached by short mesenteries to the ventral surface of the kidneys. These are the **ovaries,** the site of egg production. The ovaries vary in size according to the season. If ripe, the eggs will be black in color, with a little white on one surface. The **oviducts** are coiled tubes extending

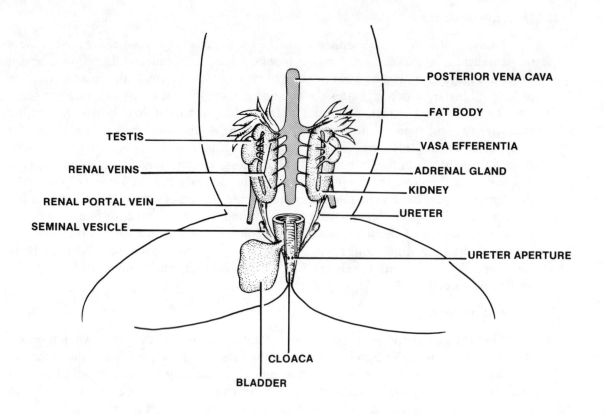

FIGURE 19.6
MALE UROGENITAL SYSTEM

POSTERIOR VENA CAVA
FAT BODY
TESTIS
VASA EFFERENTIA
RENAL VEINS
ADRENAL GLAND
KIDNEY
RENAL PORTAL VEIN
URETER
SEMINAL VESICLE
URETER APERTURE
CLOACA
BLADDER

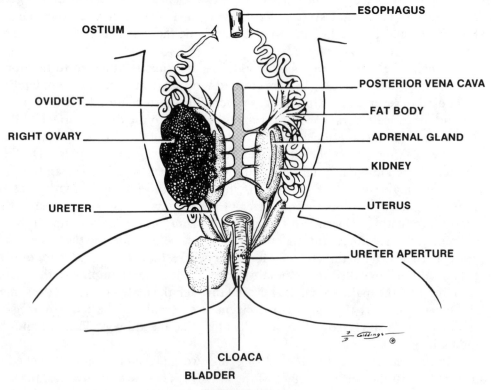

FIGURE 19.7
FEMALE UROGENITAL SYSTEM

ESOPHAGUS
OSTIUM
OVIDUCT
POSTERIOR VENA CAVA
FAT BODY
RIGHT OVARY
ADRENAL GLAND
KIDNEY
UTERUS
URETER
URETER APERTURE
CLOACA
BLADDER

from the anterior part of the coelom to the cloaca. The anterior end has a small opening, the ostium, through which eggs released into the pleuroperitoneal cavity at the time of **ovulation** enter the oviduct. In traveling down the oviduct they become covered by layers of a gelatinous substance. They may be stored temporarily in an expanded area of the oviduct near its junction with the cloaca. The storage site is known as the **ovisac** or **uterus.** The sexual embrace of the male stimulates the release of the eggs from the ovisac into the cloaca and out into the water.

Anterior to both the testes and the ovaries are **fat bodies;** these are yellowish structures with long, finger-like extensions. They function in storage of food material used during hibernation.

4. Circulatory System

This system consists of a **heart, blood vessels** (arteries, capillaries, veins), **spleen** (a small round structure on the left side of the mesentery supporting the intestine; functions in the manufacture and storage of blood cells), and **lymphatic network.** In this lab we will not make a detailed study of this system. Instead, familiarize yourself with the information given below, using Figures 19.8 and 19.9 as a guide to find the structures mentioned.

a. Heart

Carefully remove the **pericardium,** the membrane enveloping the heart, to expose the chambers of the heart and the blood vessels associated with it. Identify the cone-shaped muscular **ventricle** at the posterior end of the heart. Blood moves from the ventricle through the **conus arteriosus,** which divides into a **right** and **left aortic arch.** On either side of the conus arteriosus are two thin walled chambers, the **atria.** Lift the ventricle toward the head, and note the dark thin-walled **sinus venosus** on the dorsal side. The sinus venosus is not in Figure 19.8 or Figure 19.9 but it drains into the right atrium. Entering the posterior end of the sinus venosus is the **posterior vena cava,** and entering on either side anteriorly is a pair of smaller **anterior vena cava.**

b. Vessels

Blood entering the atria is forced into the ventricle when the atria contract. Contraction of the ventricle and conus arteriosus then forces the blood to the body and lungs. Each aortic arch has three branches: (1) the **carotid arch,** carrying blood to the head region, (2) the **systemic arch,** carrying blood to most of the body. The two systemic arches eventually fuse middorsally to form the **dorsal aorta** with its many branches leading to the internal organs, legs, etc., (3) the **pulmocutaneous arch,** carrying blood to the lungs and skin for gas exchange. All of these are classified as **arteries,** vessels which carry blood away from the heart. After blood has been transported to the tissues and the cells adjacent to the capillaries have obtained their nutrients and released their wastes, the blood returns to the heart via the type of blood vessel known as a **vein,** a vessel which carries blood toward the heart. The major veins into which blood returning from body parts drains are the **anterior vena cava** and **posterior vena cava.** Note the smaller veins leading to these two from the head, kidneys, gonads, etc.

c. Lymph

This network of vessels picks up fluid between the cells and eventually empties it back into the blood. Contractions of body muscles adjacent to large lymph vessels and actual pulsating "lymph hearts" in the lymph vessels move the lymph into the

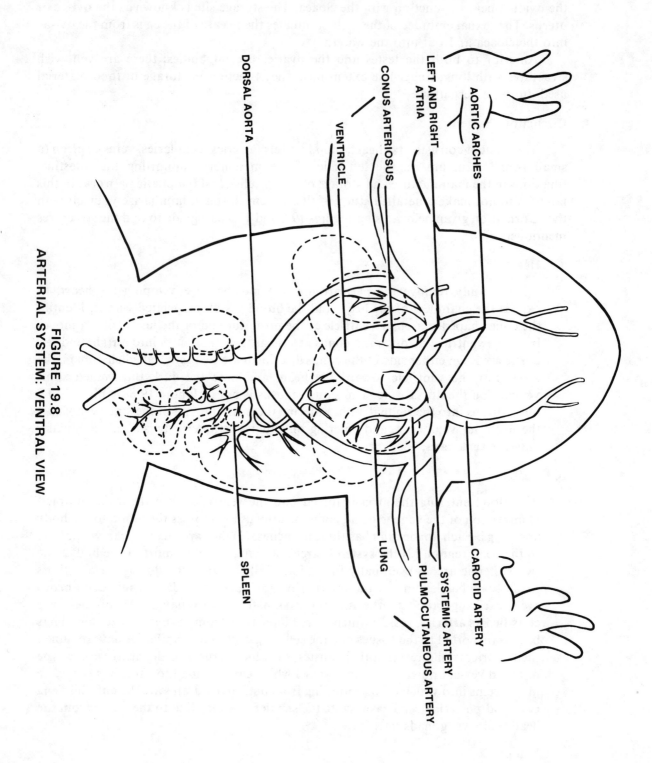

**FIGURE 19.8
ARTERIAL SYSTEM: VENTRAL VIEW**

AORTIC ARCHES

LEFT AND RIGHT
ATRIA

CONUS ARTERIOSUS

VENTRICLE

DORSAL AORTA

SPLEEN

LUNG

PULMOCUTANEOUS ARTERY

SYSTEMIC ARTERY

CAROTID ARTERY

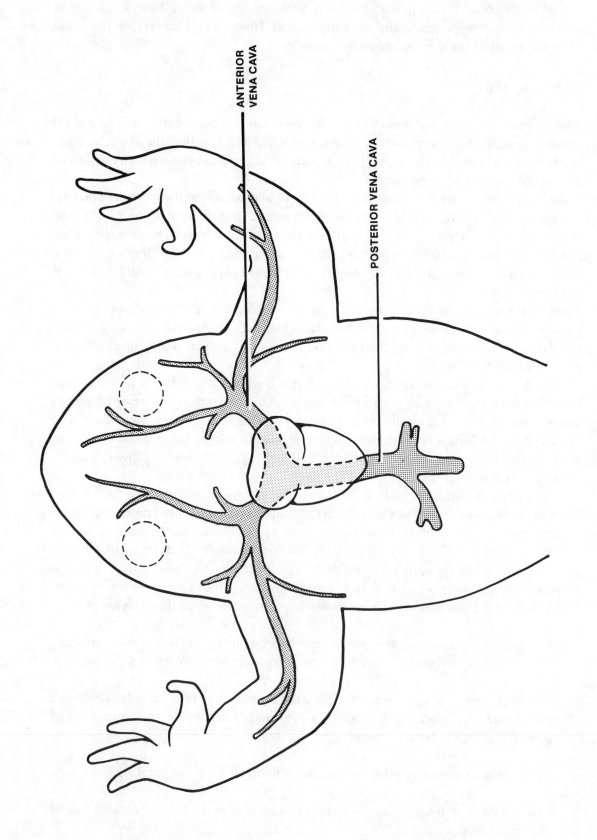

ANTERIOR
VENA CAVA

POSTERIOR VENA CAVA

FIGURE 19.9
VENOUS SYSTEM: VENTRAL VIEW

veins. Much of the lymphatic system is not very conspicuous; however, the frog does have several subcutaneous lymph sacs. These sacs account for the loose attachment of the skin to underlying muscle.

5. Nervous System

The nervous system of vertebrates can be divided into two basic parts: (a) The **central nervous system** which is composed of the five-lobed brain within the skull, and the spinal cord in the vertebral column; (b) the **peripheral nervous system** which consists of the cranial nerves and spinal nerves.

Place the frog on its back. Lift the digestive organs and kidneys and cut the mesenteries, leaving the dorsal aorta in place. Cut the intestine and urogenital ducts anterior to the cloaca and pull the entire mass of organs forward toward the head. Then cut the arches coming off the aorta and remove the organs by cutting through the posterior end of the pharynx. By removing all these internal organs, you will expose the ventral branches of the spinal nerves.

Spinal nerves leave the spinal cord and emerge between the vertebrae. Each of the 10 pairs of spinal nerves divides into two branches: an inconspicuous dorsal branch supplying muscles and skin on the back and the ventral branch supplying the ventral portion of the body wall.

Note that the second spinal nerve is the largest one of all; it forms a network with branches from the first and third spinal nerves. This network is the **brachial plexus** which supplies the muscles and skin of the shoulder and arm. The seventh, eighth, and ninth nerves also form a network called **sciatic plexus,** supplying the skin and muscles of the posterior abdomen, the pelvis, and the hind leg. Figure 19.10 will help you identify the two plexus networks.

To expose the **brain,** skin the head of the middorsal region and remove the muscles from the posterior area of the skull. Carefully chip away the bone, exposing the brain located from about midway between the eyes to the back of the skull. Continue posteriorly and expose the spinal cord by cutting through the dorsal part of each vertebra and pulling the vertebra apart. Note that the central nervous system is covered by protective membranes **(meninges).** Remove these to see the brain parts more clearly. Refer to Figure 19.10 and identify the following parts starting at the anterior end.

a. **Cerebrum** or **telencephalon** — two hemispheres with the **olfactory lobes** extending anteriorly (the first cranial nerves extend from these lobes to the nasal cavities.

b. **Diencephalon** — a small medial structure covered by a thin vascular roof. The pineal gland or **epiphysis,** is located on the dorsal surface; the optic chiasma and **pituitary gland** are on the ventral surface.

c. **Optic lobes** or **mesencephalon** — two swollen lateral lobes on the dorsal surface.

d. **Cerebellum** or **metencephalon** — a transverse band of nervous tissue located behind the optic lobes.

e. **Medulla oblongata** or **myelencephalon** — the base of the brain, continuous with the spinal cord; it has a thin vascular roof.

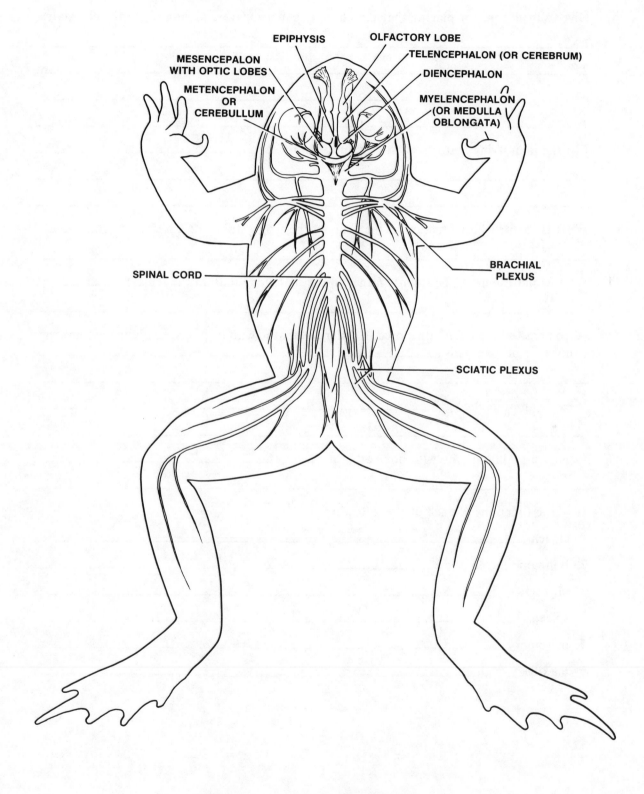

FIGURE 19.10
NERVOUS SYSTEM: DORSAL VIEW

V. Review Questions

A. Give four reasons for placing the hemichordates in a separate phylum from the chordates.

   1. _____

   2. _____

   3. _____

   4. _____

B. List the identifying features of the Phylum Chordata. _____

   _____

   _____

C. What is the probable function of the folds in the frog's stomach? _____

   _____

D. Why would the ureter be slightly larger in the male frog than in the female? _____

   _____

E. A frog has to keep its mouth closed in order to breathe with its lungs. Explain why the same is not true of mammals. _____

   _____

F. What is an amniotic egg? _____

   _____

G. Compare poikilothermy with homeothermy. _____

   _____

H. Give the functions of each of the following:

   1. cloaca _____

   2. tympanic membrane _____

   3. fat bodies _____

   4. spleen _____

   5. mesentery _____

   6. maxillary teeth _____

GENETICS — HEREDITY

I. Objectives

After the completion of this exercise, the student should be able to do each of the following:

A. Define and apply the following terms in genetics problems: **gene, trait, genotype, allele, heterozygous, homozygous, phenotype, test cross, P generation, F_1 generation, F_2 generation, monohybrid cross, dihybrid cross, dominant, recessive,** and **codominant (incompletely dominant).**

B. List the different A-B-O blood types, give the possible genotype(s) responsible for each phenotype, and solve problems involving blood type.

C. Use the **Punnett square** or the **forked line method** for determining the **genotype** and **phenotype ratios** expected from any given cross.

D. Determine, by counting corn kernels, whether the kernels were produced as a result of (a) a cross between two heterozygous plants, (b) a test cross with a heterozygous plant, (c) a testcross with a homozygous dominant plant, or (d) a cross between two homozygous recessive plants.

E. Explain the processes of DNA replication, transcription, and translation (protein synthesis).

F. Answer any of the questions, including the review questions, in this exercise.

II. Introduction

In sexual reproduction of multicellular organisms, two haploid gametes unite at fertilization to form a diploid zygote, which by successive divisions will give rise to a diploid organism. Every cell of that organism contains all the genetic information present in the original zygote with one notable exception — the reproductive cells. In the formation of those reproductive cells, **meiosis** occurs and each reproductive cell thus contains only half of the genetic information present in a diploid cell. That information is contained in the chromosomes, which occur in pairs in a diploid cell. Each chromosome consists of hundreds of structural and functional subunits called **genes.** Each gene contains coded information for making a specific protein. And it is the complement of proteins (particularly enzymes) which an individual is capable of producing that determines the individuals observable physical **traits,** or characteristics, such as eye color, height, blood type, etc.

For a trait such as eye color, each individual possesses his own **phenotype** such as blue eyes. For any one trait, then, there may be a number of different possible phenotypes.

In a diploid cell, genes occur in pairs just as do the chromosomes. A trait may be controlled by only one pair of genes, or it may be controlled by more than one pair of genes. In the following discussion and activities we will consider only traits that are controlled by one pair of genes.

The members of a **homologous** (matched) pair of chromosomes are matched gene for gene, but often the two members of a gene pair may be slightly different. Different forms of a gene that occupy the same position on homologous chromosomes are called **alleles.** It is possible for more than two different alleles to exist for a particular gene location; however, one individual could only possess two of them. If an individual possesses two different alleles for a given trait, the individual is said to be **heterozygous** for that trait; however, if both genes are alike, the individual is said to be **homozygous** for that trait. Even though the phenotype can be observed, the actual genetic makeup is not always apparent. The genetic makeup of an individual is called the **genotype.** In the case of an individual that is heterozygous for a particular trait, **both** genes may influence the resulting phenotype of that individual. When the resulting phenotype is influenced by two different alleles in the same individual, this is considered to be a type of **intermediate inheritance,** or **incomplete dominance (codominance).**

In many cases, one allele may completely mask the presence of the other one. The "stronger" allele is thus called the **dominant** gene, or allele, and the "weaker" is the **recessive.** The dominant phenotype will be present in either of the following: (1) both genes are dominant, or (2) one gene is dominant and one is recessive. The recessive characteristic will be present **only** when both genes are recessive. Inheritance of this type is said to be **dominant-recessive inheritance.**

In determining how traits are inherited, we must consider the following: Since the segregation of information during meiosis and recombination of information during fertilization is left to chance, we use laws or probability in predicting the percentage of individuals expected to exhibit certain characteristics from a given genetic cross. We need to take into consideration all possible gametes formed and all combinations of genes possible as a result of fertilization.

In this lab we will examine examples of human inheritance as well as the inheritance of certain characteristics in corn.

III.   Human Inheritance

Even though controlled genetic experiments are not conducted in order to determine human inheritance, it has been possible by tracing family histories to determine human inheritance for some physical features and physiological characteristics.

In this section of the lab, we will look at some examples of dominant-recessive inheritance in humans and at the inheritance of blood types.

A.   Dominant-recessive Inheritance

Working with a lab partner, determine whether or not you possess the following characteristics. Indicate your phenotype and your possible genotype(s) and those of your lab partner. The phenotype is expressed in words describing the characteristic possessed. The genotype is expressed in a shorthand form using letters to indicate the genes present for each trait. For a given trait, the same letter is used twice (since genes occur in pairs). In the case of dominant-recessive inheritance, a capital letter is used to indicate the presence of the dominant gene; and a small letter to indicate the presence of a recessive gene. Use a different letter for each of the eight traits that follow.

1. **PTC taster**

   The ability to taste phenyl-thio-carbamide is controlled by a dominant gene. (Get a small piece of paper saturated with the chemical and chew it; if it "tastes" just like paper, you are a non-taster.) Non-tasters have homozygous recessive genotypes.

   Your phenotype: _____   Partner's phenotype: _____

   Your possible genotype(s): _____   Partner's possible genotype(s): _____

   If you and your partner were of the opposite sex and married, list all the possible genotypes that your children might have: _____

   _____

   If your first child was a non-taster, then your genotype would have to be _____ and your partner's would have to be _____.

2. **U-shaped Tongue**

   The ability to roll the tongue into a trough-like shape when extending it from the mouth; controlled by a dominant gene.

   Your phenotype: _____   Partner's phenotype: _____

   Your possible genotype(s): _____   Partner's possible genotype(s): _____

   If you and your partner were siblings, what possible combinations of genotypes might your parents have? (List **all possible** combinations)

   _____

   _____

   If your mother was a non tongue-roller then your genotype would have to be _____ and your partner's would have to be _____.

3. **Widow's Peak**

   Hairline pointed over the center of the forehead; the gene for this is dominant over the gene for straight or curved hairline. See Figure 20.1.

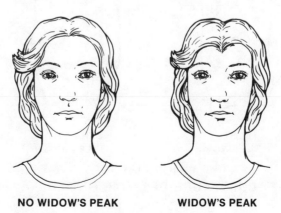

NO WIDOW'S PEAK   **WIDOW'S PEAK**

FIGURE 20.1
HAIRLINE

Your phenotype: _____ Partner's phenotype: _____

Your possible genotype(s): _____ Partner's possible genotype(s): _____

4. **Free Ear Lobe**

The lobe is not attached; controlled by a dominant gene. See Figure 20.2.

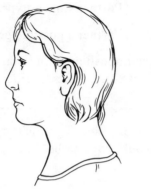

FREE EAR LOBE   **ATTACHED EAR LOBE**

FIGURE 20.2
EAR LOBES

Your phenotype: _____ Partner's phenotype: _____

Your possible genotype(s): _____ Partner's possible genotype(s): _____

5. **Bent Little Finger**

The end segment of the little finger bends toward the ring finger when placed against a straight edge; controlled by a dominant gene. If the fingers are straight and extend parallel to each other out to the very tip, this is recessive. See Figure 20.3.

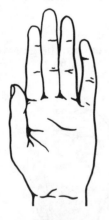 

BENT   **STRAIGHT**
LITTLE FINGER   **LITTLE FINGER**

FIGURE 20.3
CURVATURE OF LITTLE FINGER

Your phenotype: _____ Partner's phenotype: _____

Your possible genotype(s): _____ Partner's possible genotype(s): _____

6. **Finger Hair**

Presence of hair on the middle segment of the finger, dorsal surface, is controlled by a dominant gene.

Your phenotype: _____ Partner's phenotype: _____

Your possible genotype(s): _____ Partner's possible genotype(s): _____

7. **Hitch-hiker's Thumb**

The ability to bend the thumb back (without force from the other hand) so that at one point it may appear slightly parallel with the extended arm; controlled by a recessive gene. See Figure 20.4.

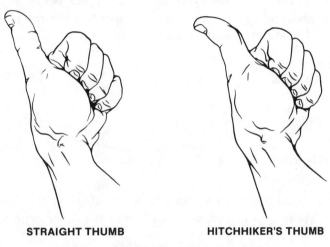

STRAIGHT THUMB　　　**HITCHHIKER'S THUMB**

FIGURE 20.4
MOBILITY OF THUMB JOINTS

Your phenotype: _____ Partner's phenotype: _____

Your possible genotype(s): _____ Partner's possible genotype(s): _____

8. **Interlocking Fingers**

When clasping the hands together such that the fingers are interlocked, note which thumb ends up on the top naturally. If the left thumb folds over the right, this is controlled by a dominant gene.

Your phenotype: _____ Partner's phenotype: _____

Your possible genotype(s): _____ Partner's possible genotype(s): _____

B. Solving Genetics Problems

In the following exercises you should keep in mind some basic facts and follow an orderly sequence of steps to prevent confusion while trying to find logical solutions.

1. The first step is to establish which symbols you will use, and then use them throughout the problem. Example: B = gene for brown eyes, b = gene for blue eyes.

2. Remember that each individual is diploid, and therefore must be shown to have two genes for each trait to be studied. Example: BB, Bb, or bb.

3. Remember that egg cells and sperm cells are produced by meiosis, so each gamete can have only one gene for each trait. Example: B or b.

4. The percentage, or fraction, of different gametes which can be expected to be produced by an individual is determined by the individual's genotype. If he is homozygous, all his gametes will have the same kind of gene. Example: BB individuals will produce only B gametes, bb individuals will produce only bb gametes.

   If the individual is heterozygous (Bb), there are two possibilities, just as there are two possibilities when flipping a coin. The chances are 1/2 that a coin will come up "heads" on a toss. The probability that a Bb individual will produce a B gamete is also 1/2.

5. Gametes recombine at random. A B sperm does not seek out B eggs. It will fertilize any egg it meets. Thus if two Bb individuals mate, random recombination of gametes would be expected to produce offspring with the genotypes below at the frequencies given, because the probability of any egg and sperm meeting is the **product** of their probabilities of being produced in the first place.

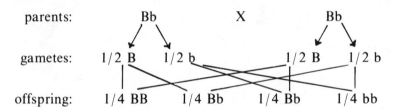

   Note that there are two ways in which Bb offspring can result from this cross, so they will be expected to occur twice as often as the other genotypes.

6. Always list the genes in the same order throughout the problem so that you will be able to recognize similar genotypes. Example: Bb is the same as bB. List the dominant gene first in problems involving dominant and recessive genes.

7. When dealing with genes for two or more traits at the same time, always group alleles for the same trait together. Example: (right) - RrTt, (wrong) - RTrt.

8. When dealing with genes for two or more traits at the same time, each gamete must be shown to contain **one of each kind of gene.** Example: an RrTt individual can produce RT, Rt, rT, and rt gametes; never Rr or Tt, or R or t gametes.

C.   A-B-O Blood Types

   Different blood types result from differences in specific proteins in the blood plasma and on the surface of the red blood cells. The proteins in the plasma of the blood are called **antibodies.** These substances act upon certain proteins present on the red blood cells called **antigens,** causing the cells to clump together. (In a few weeks you will be doing a lab on circulation in which you will determine your blood type.) Antibody A acts upon antigen A and so the two are never present together; the same is true for antibody B and antigen B.

   The blood types are transmitted from parent to offspring by a series of three allelic genes. These are: a gene for the A antigen, designated I^A; a gene for the B antigen, I^B; and a gene for the absence of antigen, i. I^A and I^B are dominant over i, but are codominant with respect to each other. Thus, a person carrying $I^A I^B$ genes has both antigens and is of the blood type AB. The possible genotypes for each blood type are given in Table 1.

TABLE I

BLOOD TYPE (Phenotype)	GENOTYPE	ANTIGENS ON THE RED BLOOD CELLS	ANTIBODIES IN THE PLASMA
O	ii	none	A and B
A	$I^A i$, $I^A I^A$	A	B
B	$I^B i$, $I^B I^B$	B	A
AB	$I^A I^B$	A and B	none

1. If a woman with type O blood married a man with type AB blood, represent the following:

   Wife's genotype: _____   Husband's genotype: _____

   Genetic make-up
   of egg cells: Genetic make-up
                                          of sperm cells:

   Possible combinations (i.e., possible genotypes of their offspring):

   _____

2. Assume that a woman with type AB blood marries a man with type AB blood and represent the following:

   Wife's genotype: _____   Husband's genotype: _____

   Genetic make-up
   of egg cells: Genetic make-up
                                          of sperm cells:

   Use a Punnett square to show possible combinations:

   EGGS

   SPERM

   What are the chances (express in fraction form) that their first child will have type AB blood? _____ type A? _____ type B? _____ type O? _____

IV. Inheritance in Maize (Indian Corn)

A. **Monohybrid Cross**

   A strain of corn producing pure purple kernels (PP) is crossed with a strain producing pure yellow kernels (pp). Purple is dominant with the resulting F_1 ears all bear purple kernels. When the F_1 is self-pollinated, the resulting F_2 ears bear both purple and yellow kernels. The dominant allele for kernel color is _____ and the recessive allele for kernel color is _____.

Count the number of purple and yellow kernels on one row of the F_2 ear without removing the kernels. Determine the ratio of purple to yellow. Now tabulate below the numbers obtained by each student (or row of students) and add these figures to get a total. Using the total numbers, determine a ratio of purple to yellow.

PURPLE	YELLOW	RATIO OF PURPLE TO YELLOW
_____	_____	_____ : _____
_____	_____	_____ : _____
_____	_____	_____ : _____
_____	_____	_____ : _____
_____	_____	_____ : _____
_____	_____	_____ : _____
TOTAL _____	_____	_____ : _____

In order to determine the expected ratio of such a cross, it is necessary to first indicate the parent (P) generation:

PP   X   pp

Under the genotype of each parent indicate the genetic make-up of the gametes that the parent may produce. If all the eggs come from one parent and all the sperm nuclei from the other in order to produce the F_1 generation, then every plant in the F_1 generation would have what genotype? _____

Next, show a cross between two individuals in the F_1 generation and indicate the genetic make-up of possible gametes from each.

F_1 generation _____ X _____

Fill in the Punnett square to show the possible genotypes obtained by this cross:

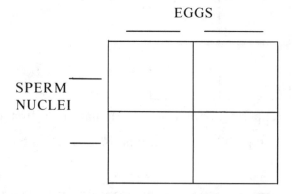

What is the expected phenotypic ratio of the F_2 generation? _____

B. Test Cross or Back Cross

To determine the genotype of an organism which exhibits the dominant feature (phenotype), a **test cross** is used. A test cross always involves a homozygous recessive and the unknown genotype. A strain producing yellow kernels (pp) is crossed with a strain of unknown genotype but with obviously purple phenotype.

Count the number of purple and yellow kernels on one row and determine the ratios and total, using numbers obtained from the other students or rows.

PURPLE	YELLOW	RATIO OF PURPLE TO YELLOW
_____	_____	_____ : _____
_____	_____	_____ : _____
_____	_____	_____ : _____
_____	_____	_____ : _____
_____	_____	_____ : _____
_____	_____	_____ : _____
TOTAL _____	_____	_____ : _____

Show the two possible outcomes of a test cross with a purple plant:

P generation _____ X _____ or _____ X _____

Possible gametes:

C. The Dichotomous or Forked-Line Method

The Punnett square becomes especially time and space consuming when one is dealing with dihybrids or greater. So the **dichotomous** or **forked-line method** is usually utilized to find the possible results of crosses.

When the answer calls for only phenotypic results, then simply write the expected phenotypic ratio of the cross of one trait and multiply each of these probabilities by the expected phenotopic ratio of the other trait(s). In other words, a dihybrid is nothing more than 2 monohybrids being considered at the same time.

For example, the F_1 heterozygous purple, starchy corn (PpSs) are crossed with each other. The results will probably be: 3/4 purple, 1/4 yellow, 3/4 starch, and 1/4 sweet. Using the forked-line method, one should write:

```
                  3/4 starchy = 9/16 purple starchy
   3/4 purple
                  1/4 sweet = 3/16 purple sweet

                  3/4 starchy = 3/16 yellow starchy
   1/4 yellow
                  1/4 sweet = 1/16 yellow sweet
```

Thus the 9:3:3:1 ratio is derived without the use of the laborious Punnett square. Also one has a better concept of Mendelian genetics —that is, two genes governing two traits usually assort independently from each other during meiosis.

If genotypic results are needed, then write the expected genotypic ratio of one trait's cross and multiply each of these probabilities by the genotypic ratio of the other trait's cross.

For example, the F_1 heterozygous purple, starchy corn (PpSs) are self-pollinated.

```
              1/4 SS = 1/16 PPSS
   1/4 PP     2/4 Ss = 2/16 PPSs
              1/4 SS = 1/16 PPss

              1/4 SS = 2/16 PpSS
   2/4 Pp     2/4 Ss = 4/16 PpSs
              1/4 ss = 2/16 Ppss

              1/4 SS = 1/4 ppSS
   1/4 pp     2/4 Ss = 2/16 ppSs
              1/4 ss = 1/16 ppss
```

It is doubtful that this 1:2:1:2:4:2:1:2:1 ratio is important for most problems. Some questions may ask for one or more of these probabilities such as what is the chance of an F_2 organism being PpSs?

Also with this method, trihybrids, back-crosses, and incomplete dominance are easily incorporated into the results.

For example, what is the probability that a woman (homozygous for PTC non-tasting, heterozygous for free-ear lobe, and blood type AB) and a man (heterozygous for PTC tasting, heterozygous for free-ear lobe, and blood type O) will have a baby boy who tastes PTC, has a free-ear lobe, and has A blood type?

Answer: Take each trait separately.

Woman tt (non-taster)   X   man Tt (taster) = 1/2 Tt, 1/2 tt

    Ee (free-ear lobe)   X   Ee (free-ear lobe) = 1/4 EE, 1/2 Ee, 1/4 ee

∴ 3/4 free-ear lobe, 1/4 attached

$I^A I^B$ (AB blood)   X   ii (O blood) = 1/2 I^Ai, 1/2 I^Bi

Chromosomes XX (female)   X   XY (male) = 1/2 female, 1/2 male

Thus, the probability of a boy who tastes PTC, has a free-ear lobe, and A blood type is

$$1/2 \times 3/4 \times 1/2 \times 1/2 = 3/32$$

D. Dihybrid Cross

A pure strain of corn producing purple-starchy kernels (PPSS) is crossed with a pure strain producing yellow-sweet (ppss). The resulting F_1 ears all bear purple-starchy kernels. When the F_1 is self-pollinated, the resulting F_2 generation contains various combinations.

Carefully count the number of kernels of each phenotype appearing on a row of F_2 ear. Tabulate the results as done in preceding exercises.

PURPLE & STARCHY	PURPLE & SWEET	YELLOW & STARCHY	YELLOW & SWEET	RATIO
_____	_____	_____	_____	_____
_____	_____	_____	_____	_____
_____	_____	_____	_____	_____
_____	_____	_____	_____	_____
_____	_____	_____	_____	_____
_____	_____	_____	_____	_____

TOTAL _____   _____   _____   _____   _____

How many phenotypes are in the F_2 generation? _____

Which two phenotypes are new combinations? _____

Indicate each of the following:

P generation _____ X _____

Possible gametes:

F_1 generation _____ X _____

Find the genotypic and phenotypic ratios expected in the offspring of this cross using the dichotomous or forked line method.

V.   DNA and the Genetic Code

A model of **DNA (deoxyribonucleic acid)** was produced in 1953 by Francis Crick and James Watson. Their model of a DNA molecule appeared as a spiralling ladder. The sides were composed of alternating **deoxyribose sugar** and **phosphate** molecules and the rungs were made of 2 of 4 **nitrogen bases**.

Actually, the DNA molecule is a large polymer, composed of smaller monomeric units known as **nucleotides.** Each nucleotide consists of a deoxyribose sugar, a phosphate, and either a **purine (adenine** or **guanine)** or a **pyrimidine (thymine** or **cytosine).** By dehydration synthesis the nucleotides are linked together to form strands. These strands run in opposite directions, or are **anti-parallel.** The strands are connected by 2 **hydrogen bonds** between adenine and thymine or 3 hydrogen bonds between guanine and cytosine.

1. Cut out the models on page 203. Tape a phosphate, a sugar, and a nitrogen base together to form a nucleotide. Then fit them together to form a section of a DNA molecule. Do not tape together. Note (1) the hydrogen bonds (2) the 5' carbon atom and 3' carbon of the sugar connect the phosphates and give the strand direction (3) 1' carbon of the sugar attaches to the nitrogen base.

2. Arrange the DNA section so one strand's nitrogen bases "read" G,A,T, and G from a 3' to 5' direction.

VI.   DNA Replication

Prior to cell division (mitosis or meiosis), DNA replicates to form two identical molecules. After the separation of the strands in a section of the DNA, complementary nucleotides are added to each strand by an enzyme known as **DNA Polymerase III.** Eventually, all sections of the molecule are replicated. Thus, each strand serves as a template to form a molecule identical to the original.

3. Together with a partner, recreate the DNA section in the above exercise. Now separate the strands. Then, add complementary nucleotides one at a time until two complete molecules are produced.

VII.   RNA Transcription

RNA (ribonucleic acid) is produced by using the template message of DNA. This process is similar to replication; however, RNA is produced and only 1 strand of DNA serves as the template. This strand is often known as the **sense** strand.

A section of DNA uncoils as **RNA polymerase** connects itself to the sense strand. Proceeding in a 5' to 3' direction of that DNA strand, the enzyme adds complementary nucleotides. These nucleotides differ from those of DNA. Thymine is absent but **uracil** is used instead as a complementary base to adenine. Also the pentose is **ribose,** not deoxyribose.

The finished product is a single-stranded RNA molecule. This molecule will be modified by other enzymes to become **transfer RNA, messenger RNA,** or **ribosomal RNA.**

4. Cut out the models of ribose, phosphate, and nitrogen bases. Arrange these into nucleotides and tape.

   With a lab partner use the models from the previous exercises to create a section of DNA molecule in which one strand reads in a 5' to 3' direction the following sequence: T A C C G A G T.

   Now separate the two strands of DNA to approximately a foot apart. Transcription has begun!

   Match up complementary RNA nucleotides on the strand which reads T A C C G A G T in a 5' to 3' direction. Remove this RNA strand from its DNA complement and reattach the DNA "sense" strand to the other DNA strand. This RNA chain represents what will become a transfer, messenger, or ribosomal RNA.

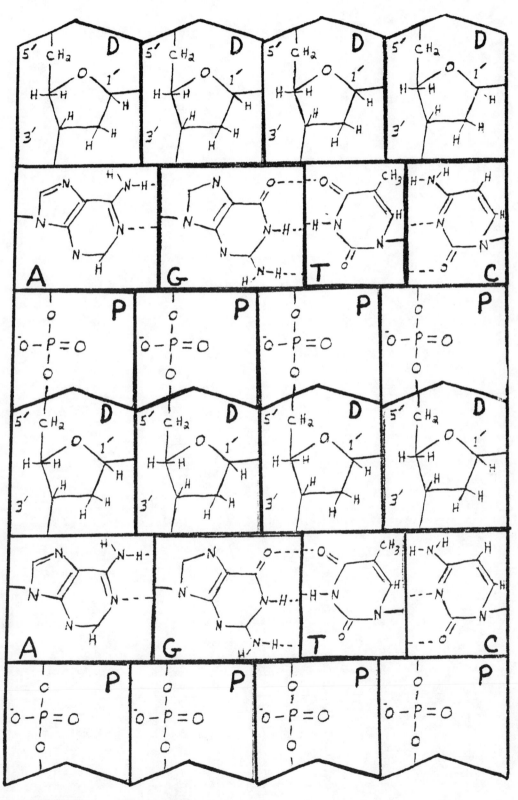

A = adenine   C = cytosine
G = guanine   D = deoxyribose
T = thymine   P = phosphate

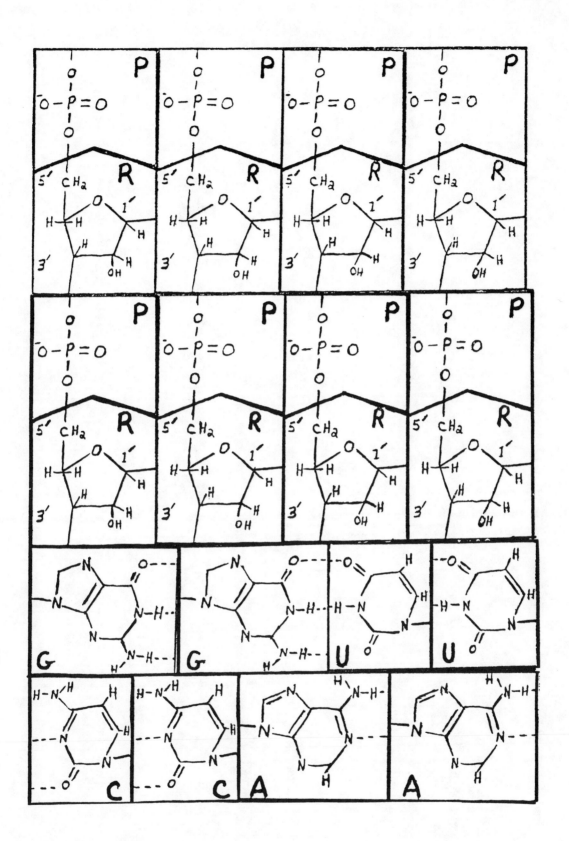

VIII. Translation or Protein Synthesis

The process by which RNA codes for the assembly of protein is known as **translation.** All three types of RNA are involved. Two **ribosomal subunits, large** and **small,** are each composed of protein and ribosomal RNA.

The synthesis of protein begins with the attachment of the small ribosomal subunit to a particular area of a messenger RNA (mRNA). This area consist of an **initiating codon,** three nucleotides which have the nitrogen bases adenine, uracil, and guanine in a 5' to 3' direction. The small ribosomal subunit also extends past to one additional codon or 3 nucleotide set. Then the large ribosomal subunit joins the mRNA and the small ribosomal subunit complex.

A transfer RNA (tRNA), charged with one of the 20 different amino acids, unites by hydrogen bonding to the initiating codon. Another amino acid-charged tRNA bonds to the adjacent codon. The 2 amino acids are now close enough spatially to be bonded together by an enzyme. This resulting covalent bond is a peptide bond and the reaction releases water (dehydration synthesis).

The tRNA on the 5' side of the ribosome now is released and the ribosomal unit moves 3 nucleotides or 1 codon toward the 3' of the mRNA.

Another charged tRNA brings its amino acid to the ribosome and unites its anticodon to the next codon of the mRNA. Again dehydration synthesis occurs to bond the next amino acid to the growing peptide chain.

This process continues until one or more **terminating codons** are encountered. For example, UAG on the mRNA codes for a tRNA which possesses no amino acid. Thus, the chain cannot add any more amino acid and the polypeptide is complete.

Protein results when one or more polypeptides are somewhat modified by other enzymes and then united together.

IX. Review Questions

A.   If the father has type A blood, the mother has type B blood, and the child has type O blood; indicate each of the following:

Father's genotype: _____ Mother's genotype: _____ Child's genotype: _____

B.   Explain why the homozygous recessive would always be used in a test cross.

C.   If a human male and female produce children, what proportion of their offspring would be males? What proportion would be females? Illustrate, using a Punnett square.

D.   In rabbits the gene for white fat (W) is dominant over the gene for yellow fat (w). Suppose that a pure white-fatted male rabbit is mated to a pure yellow-fatted female rabbit.

   1.   What color would all the first generation (F₁) offspring be? Why?

   2.   What is the genotype of the female parent? The male parent?

   3.   What is the genotype of the offspring?

   4.   What proportion of the eggs would contain genes for white-fattedness in the female parent? Why?

5. What proportion of the gametes of the F_1 generation will have genes for:

Yellow-fattedness only?

White-fattedness only?

White and yellow-fattedness in equal proportions?

E. Long wings in Drosophila are due to a dominant gene (V) and vestigial wings to its recessive allele (v). A vestigial winged male and a heterozygous long-winged female are mated. What types of gametes can be produced by the vestigial winged male? What type of gametes can be produced by the heterozygous long-winged female? What types of offspring can be produced and in what proportion? Use a Punnett square to explain.

F. A blue-eyed man whose parents were brown-eyed marries a brown-eyed woman whose father was brown-eyed and whose mother was blue-eyed. If they have children, what will the probability of their being blue-eyed? brown-eyed?

G. In guinea pigs short hair is dependent upon the dominant gene (S) while long hair is due to the allele (s). Rough coat is due to the dominant gene (R) and smooth coat to its recessive (r). Black is dependent upon the dominant gene (B) while white is dependent upon the allele (b). A guinea pig of genotype SSRRBB was crossed with one of the genotype ssrrbb. What was the appearance of the F_1 generation?

Two F_1 individuals were crossed many times (ie, many offspring), what phenotypic ratio should be expected in the F_2 generation?

H. Black Langshan chickens have feathered shanks while Buff Rocks have unfeathered shanks. When these two breeds are crossed the F_1 have feathered shanks. The feathered condition is due to the presence of two pairs of genes, F,f and S,s where F is dominant to both f and s. The same is true for S which is also dominant to f and s. The feathered condition is produced by the presence of one or more dominant genes. The homozygous recessive condition of f and s produces the unfeathered shanks. What would be the expected phenotypic ratio in the F_2 if the two F_1 individuals from the Langshan and Rock parents are crossed?

I. Normal vision (C) in man is dominant to colorblindness (c) and is sex-linked A normal-visioned man, whose father was colorblind marries a colorblind woman. What are the chances that a son will be colorblind? A daughter? Explain.

J. The determiner for brown-eyes (B) is dominant to blue-eyes (b) and is not sex linked. A colorblind man with brown eyes, whose mother was blue-eyed, marries a normal visioned blue-eyed woman, whose father was colorblind. Show the expected phenotype ratio of their children involving eye color and colorblindness.

K. A man that has blood type A negative, whose father was B positive and whose mother was AB positive; married a woman that was B positive. The woman's mother was A negative and her father was AB positive.

1. Give the genotypes of all the above mentioned individuals.

2. Give the genotypes and phenotypes of the F_1 generation.

L. A hemophiliac man with blue eyes married a woman that was a carrier of hemophilia and had brown eyes. The woman's mother had blue eyes.

1. Give the genotypes of all the above mentioned individuals.

2. Give the genotypes and phenotypes of the F_1 generation.

3. What are this couples chances of having a normal blue-eyed son?

208

TISSUES, ORGANS AND SYSTEMS

I. Objectives

After the completion of this exercise, the student should be able to do each of the following:

A. Identify each of the following tissues and structures upon microscopic examination:

epithelium
    simple squamous
    stratified squamous
    cuboidal
    columnar
    pseudostratified ciliated columnar

connective tissue
    areolar
    adipose

goblet cells
fibroblasts
collagen fibers
basement membrane
elastic fibers

B. List the major types of tissue found in animals and describe the unique structural characteristics of each.

C. State the major function(s) of each type of tissue.

D. Relate the structure of a tissue to its function.

E. Identify the four layers of tissue found in a cross section of small intestine.

F. Answer the review questions at the end of this exercise.

II. Cellular Organization

The organization of multicellular organisms is very complex. However, the structure and function of each living thing is dependent on the structures and functions of its individual cells and the material which those cells produce. The cell, then, is the basic unit of structure and function in all living things.

The degree of complexity of cellular organization varies among animals, ranging from single-celled organisms to those composed of billions of different kinds of cells — cells which are organized into the following biological hierarchy:

1. A group of similar cells together performing a specific function, such as contraction, comprise a **tissue.**

2. An **organ** is a group of different tissues organized into a complex structural and functional unit.

3. A group of organs performing related functions are collectively an **organ system.**

4. The **organism,** the entire living thing, is composed of all its organ systems working together, among higher animals.

In this laboratory period we will study cells, tissues, and one organ — the intestine. All of the material will be that of vertebrates, the most highly organized of all animals.

III. Cells

The only cells which exist singly rather than as parts of a tissue in vertebrates are the sperm cells and egg cells.

A. Sperm Cells

Obtain prepared slides of stained sperm cells of guinea pigs, frogs, and humans and examine under high power.

What is their shape and size? _____

_____

_____

How do they seem to be structurally fitted to perform their function? _____

_____

B. Egg Cells

Observe the demonstration of egg cells available in the laboratory. Compare the size of the egg cell to that of the sperm cells.

Given equal amounts of cytoplasm, estimate the relative number of eggs produced as

compared to the number of sperm. _____

The sperm cell and egg cell both contribute material to the zygote, the first cell of a new organism.

What is contributed equally by both? _____

What is contributed almost entirely by the egg cell? _____

210

IV. Tissues

Cells become specialized, or differentiated, to perform specific functions. In multicellular organisms, groups of cells may become associated to perform a particular task. A group of similar cells involved in the same task is called a **tissue.**

Tissues are generally classified into one of four major types: **epithelial, connective, muscular,** and **nervous.** They resemble one another to the extent that each is composed of cells and intercellular material. There are many different subtypes, each with different functional modifications. The first two types of tissue will be studied in detail now; the last two will be studied in detail in later exercises. Use Figure 21.1 while examining the slides specified in this exercise.

A. Epithelial Tissue

The surface of the body, the coverings and linings of most organs, the secretory glands, and the sensory areas of sense organs are all composed of epithelial tissue. It may have one or more different functions including protection, absorption, secretion, and sensation. Structurally, it consists of cells closely packed together with very little, if any, intercellular (between cells) spaces or substance. Epithelial tissue which covers internal and external surfaces has a free surface and a basal surface in contact with other tissues, usually connective tissue. The **basement membrane,** made up of intercellular material and located at the basal surface, holds the epithelial tissue to underlying connective tissue. Epithelial tissue is subdivided into several categories based on the shape and arrangement of its cells.

1. Squamous Epithelium

This subtype consists of flat, pancake-like cells. If the cells are arranged in a single layer, the tissue is called **simple squamous epithelium.** These epithelia are unlikely to repair themselves readily. The peritoneum, pleural membranes, and linings of blood vessels consist of simple squamous epithelium. Several layers of flat cells are referred to as **stratified squamous epithelium.**
Stratified squamous epithelium is protective, guarding against dessication through loss of water. If the epithelial surface is in contact with air, its cells usually produce keratin, a substance which helps reduce water loss. If the epithelial surface is constantly bathed with fluid, dessication is not a problem, and keratin is not produced. Stratified squamous epithelium is found at the outer surface of the skin, or epidermis. Examine prepared slides of simple and stratified squamous epithelium. Remember that you have already examined the squamous epithelium of your mouth in Exercise 1.

2. Cuboidal Epithelium

These cube-shaped cells with centrally located nuclei are found in various glands of the body, such as the thyroid, pancreas, and salivary glands. They function in secretion. Obtain a prepared slide containing a cross section of the thyroid gland or the kidney and locate the **cuboidal epithelium** lining the tubules.

3. Columnar Epithelium

This tissue consists of long, column-like cells with nuclei usually located at the bases of the cells. The linings of the stomach, intestine, and other parts of the digestive system are of this type. Their function is absorption. Some flask-like cells, called **goblet cells,** are seen interspersed among typical columnar cells. They contain mucus, which is poured out over the surface of this tissue.

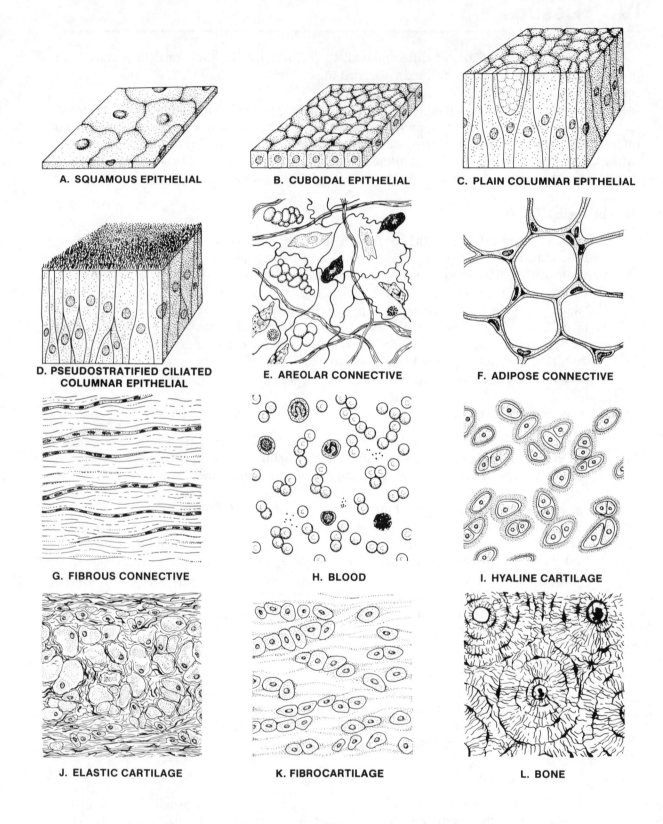

A. SQUAMOUS EPITHELIAL

B. CUBOIDAL EPITHELIAL

C. PLAIN COLUMNAR EPITHELIAL

D. PSEUDOSTRATIFIED CILIATED COLUMNAR EPITHELIAL

E. AREOLAR CONNECTIVE

F. ADIPOSE CONNECTIVE

G. FIBROUS CONNECTIVE

H. BLOOD

I. HYALINE CARTILAGE

J. ELASTIC CARTILAGE

K. FIBROCARTILAGE

L. BONE

FIGURE 21.1
VERTEBRATE TISSUES

Some columnar epithelium has delicate hair-like structures called cilia projecting from the cells at their exposed surfaces. The trachea, bronchial tubes, and Fallopian tubes are lined with **ciliated columnar epithelium.** Wave-like movements of the cilia in the respiratory tract move a sheet of mucus upward, trapping bacteria and dust particles which enter in the air. *Obtain a prepared slide* of a cross section through the frog intestine and identify the columnar epithelium, goblet cells, and basement membrane.

4. Pseudostratified Ciliated Columnar Epithelium

The cell nucleus may lie at any position within the cell in this type of epithelium, giving the tissue a layered appearance. However, all cells are actually in contact with the basement membrane.

Examine a prepared slide of a section of human trachea and identify the **pseudostratified ciliated columnar epithelium** lining the central cavity.

B. Connective Tissue

Connective tissue gives support to various organs, fills up spaces in the body, attaches organs to each other, and affords protection. The tissues of this type are characterized by cells which are interspersed in an abundance of nonliving intercellular substance termed the **matrix.** The nature of this intercellular material determines the type of connective tissue. The cells responsible for the production of the matrix are often called **fibroblasts,** since the matrix is usually composed of some fibers, primarily tough **collagenous fibers** which are thick, non-branching, and inelastic, and/or **elastic fibers,** which are thin and branching. Connective tissues are classified according to the predominant kinds of cells and/or fibers in addition to the nature of the ground substance. The ground substance may include liquid (such as in blood), a gel or semisolid substance (as in cartilage), or a solid composed of deposited minerals (as in bone).

1. Connective Tissue Proper

The intercellular matrix of this type of connective tissue always contains numerous fibers, which are composed of different types of protein. Connective tissue proper is very variable, generally containing several kinds of cells: fibroblasts secrete proteins; macrophages, common near blood vessels, become active when there is an inflammation; fat cells are highly specialized for fat storage. Tissue fluid is part of the connective tissue. It is derived from blood.

a. **Loose connective tissue** is characterized by the loose, irregular arrangements of its fibers, so that the tissue is fragile and easily torn.

    i. Areolar Connective Tissue

    This is characterized by the presence of both collagenous (thick) and elastic (thin) fibers between the fibroblasts. This type of tissue holds the skin to the body, and lends strength to walls of blood vessels and to many other organs. *Examine a slide* of **areolar connective tissue** and identify the fibroblasts, collagen fibers, and elastic fibers.

    ii. Adipose Connective Tissue

    Fibroblasts which have large vacuoles of fat are found in this type of connective tissue. These cells are also called **fat cells.** In mature cells the fat vacuole fills up most of the cell. **Adipose tissue** lies under the skin, around the kidneys, in the orbit of the eye behind the eyeball, and in a number of other locations in the body where insulation or cushioning of delicate organs is needed. *Examine a prepared slide* of adipose tissue and identify the nucleus, vacuole, and cell membrane of a single fat cell.

b. **Dense,** or **fibrous, connective tissue** is characterized by fibers which fill up most of the intercellular spaces. This makes it stronger than loose connective tissue. In this type of tissue, which composes the tough **ligaments** and **tendons,** the fibroblasts are arranged in linear rows with masses of collagenous fibers packed between them. Tendons and ligaments will be studied further in Exercise 26.

2. Blood and Lymph

Blood and **lymph** have a liquid rather than a solid matrix. They are classified as connective tissues because they contain several types of formed elements which are derived from connective tissue structures. These include **erythrocytes,** or red blood cells; **leukocytes,** or white blood cells, and; **thrombocytes,** or platelets. Blood will be examined more thoroughly in Exercise 23.

3. Cartilage

Cartilage is a specialized type of connective tissue characterized by a firmer type of intercellular material which gives it greater rigidity. **Chondrocytes** (cartilage producing cells) is the name applied to the fibroblasts of this connective tissue. They lie in cavities in the matrix called **lacunae.**

a. Hyaline Cartilage

This glistening white material is found covering the ends of bones, reinforcing the nose, and attaching the ribs to the sternum. It is also seen in the larynx, trachea, and bronchial tubes. Depending upon where it is found, it may be referred to as articular, costal, skeletal, or embryonal cartilage.

b. Elastic Cartilage

This is somewhat similar in appearance to hyaline cartilage, except that numerous elastic fibers are visible between the cells. This tissue is found in the external ear, Eustachian tube, and parts of the larynx.

c. Fibrocartilage

Fibrocartilage resembles fibrous connective tissue except that the cells are not arranged in rows. Collagenous fibers are densely packed between the chondrocytes. The intervertebral discs which form cushions between the vertebrae are made of fibrocartilage, as is the pubic symphysis which connects the pelvic bones.
Cartilage will be studied later in more detail in Exercise 26.

4. Bone

In this tissue, the intercellular matrix contains numerous fibers and water. It is also characterized by the impressive amount of inorganic salts, such as calcium carbonate ($CaCO_3$) and calcium phosphate ($Ca_3(PO_4)_2$) deposited in the matrix. An extensive study of bone will be done in Exercise 26.

C. Muscular Tissue

Muscular tissues contract to cause movement. Three types are **skeletal muscle, smooth muscle,** and **cardiac muscle.** Skeletal (usually voluntary) muscle is the type found attached to bones; smooth, or visceral (usually involuntary) muscle is found in the walls of tubular organs; cardiac muscle is restricted to the heart.
The different types of muscle cells (also called muscle fibers) vary in structure and arrangement. Both skeletal muscle and cardiac muscle cells have **striations,** or visible

crossbands in the cell due to the arrangement of internal contractile elements, whereas smooth muscle cells lack striations. Skeletal muscle cells are larger, especially in length, than the other two types and contain many nuclei. In contrast, only one nucleus is found in each cell in the other two types of muscle. Each skeletal muscle cell extends the length of the whole muscle and attaches to tendons at either end. Smooth muscle cells overlap each other forming sheets of muscle. Cardiac muscle cells are connected end to end; thick bands called **intercalated discs** indicate the connections between cells. Muscle tissue will be discussed in detail in Exercise 27.

D.   Nervous Tissue

These tissues are specialized for the reception of stimuli and the transmission, interpretation, and coordination of nervous impulses within the nervous system. Refer to your textbook for a diagram of a nerve cell **(neuron).**

Bundles of nerve fibers are called **nerves.** The cell bodies of neurons are usually located in **ganglia,** the spinal cord, or the brain. Nervous tissue often includes accessory cells as well as the neurons. Nerves will be discussed in greater detail in Exercise 24.

V.   An Organ

Some organs in the body are composed largely of one type of cell. For example, the brain is made up largely of neurons, and muscles are composed mostly of muscle cells. However, many organs are composites of many different types of tissues. Two good examples are the intestine, discussed below, and the skin, which will be covered in Exercise 25.

Study a prepared slide of a cross section of small intestine and identify the following layers of tissue from the central cavity outward.

Mucosa. This tissue is made up of simple columnar epithelium that lines the central cavity and contains glandular epithelial cells which secrete mucus. This layer is arranged in deep folds called **villi** to increase the absorptive surface area of the intestine. Within the crevices of the villi are **digestive glands** which secrete digestive juices. In some places, the outermost part of the mucosa consists of a thin muscle layer.

Submucosa. The digestive glands extend down into this layer, which is a broad band of loose connective tissue containing blood vessels and nerves.

Muscularis externa. Two distinct layers of smooth muscle make up this layer. The first is an inner circular layer of cells in which long axes of the fibers encircle the intestine. The second is an outer longitudinal layer in which the muscle fibers run parallel with the long axis of the intestine.

Serosa. This is a very thin outermost layer of simple squamous epithelium. It is often difficult to distinguish between the serosa and the outer boundary of the muscularis externa.

VI.   Review Questions

A.   How does a tissue differ from an organ? _____

B.   In what main way does connective tissue differ in structure from the other three types?

C. How do collagenous fibers differ from elastic fibers? _____
    _____

D. List the specific tissues found in the small intestine. _____
    _____

E. List three types of matrix found in connective tissues and give an example of a connective
    tissue displaying each type. _____
    _____

F. Describe three types of epithelium and give the functions of each. _____
    _____

FETAL PIG DISSECTION

exercise 22

I. Objectives

After the completion of this exercise, the student should be able to do each of the following:

A. Locate and identify the anatomical parts of the fetal pig given in boldface in this exercise, and give the function of each.

B. Define the terms of relative position and direction given in the introduction, and locate these positions and directions on any vertebrate.

C. Distinguish between male and female fetal pigs on the basis of their external features.

D. Locate the parts of the respiratory system seen in the fetal pig on the human torso model and anatomy wall chart.

E. Locate as many as possible of the parts of the digestive system given in boldface in this exercise on the charts and models which show the human digestive system.

F. Locate the parts of the heart given in boldface in this exercise on the sheep heart and on the human torso model.

G. Answer the review questions at the end of this exercise.

II. Introduction

In this exercise, the anatomy of the fetal pig, *Sus scrofa,* will be studied. This will aid us in obtaining some knowledge of mammalian anatomy and physiology. You will also examine certain comparative aspects of the anatomy and physiology of man.

The animals that you will use were removed from their mothers which were slaughtered for food. Farmers market pregnant females because hogs are sold by the pound.

In the pig, fertilization of the egg occurs in the oviduct; by the time the early embryo reaches what is equivalent to the blastula, it becomes buried in the uterine wall where subsequent development takes place. Here not only do the extraembryonic membranes form, but also a new organ, the **placenta.** This organ is made up of the **extraembryonic membranes:** amnion, chorion, and the allantois, which become intricately intermingled with the lining of the uterus. Remember that at no time does the blood of the fetus (unborn) mix with that of the mother. Gases and small molecules are able to pass across capillary walls in the placenta. The fetus lives like a parasite on the mother, absorbing all of its nourishment and oxygen from and excreting all of its wastes into the blood of the mother, all by way of placental circulation. The connection between the placenta and the fetus is through the **umbilical cord.** The period or length of pregnancy (gestation) in pigs is approximately 17 weeks. The fetal pigs used in lab will be within one or two weeks of birth.

Below is a list of anatomical terms which are used frequently in dissection directions. Familiarize yourself with the definitions before continuing this exercise.

Dorsal —	near or toward the back.
Ventral —	near or toward the belly.
Medial —	near or toward the middle.
Lateral —	near or toward the sides.
Anterior —	near or toward the head end.
Posterior —	near or toward the tail end.
Caudal —	referring to the tail or tail end.
Cephalic —	referring to the head or head end.
Longitudinal —	in the axis from head to tail.
Transverse —	a thin section which cuts across the body at a right angle to the long axis.
Superficial —	on or near the surface.
Pectoral —	relating to the chest or shoulder region.
Pelvic —	relating to the hip region.
Distal —	free end of a limb or projection or toward this free end of a limb or projection.
Proximal—	end attached to the body, or toward the end attached to the body.

The fetal pigs have been preserved in either formalin or isopropyl alcohol. These preservatives tend to dehydrate your fingertips; therefore, students may desire to lightly grease their fingertips with petroleum jelly.

The instructor will make available one fetal pig for each two students. The fetal pigs will be evenly distributed in reference to male and female. Upon receiving a fetal pig, wash it in running water and place it in a dissecting pan for observation. The aborted pigs will usually have a slash on one side of the neck. This marks the location where the blood was drained from the pig and where red and blue liquid latex were injected into a major artery or vein, respectively. The latex has become solid and rubbery in texture in the fetal pig. This strengthens the blood vessels and aids in their identification.

Measure your pig from snout to anus and refer to the table below to determine its approximate age. Remember that 1 cm. = 10 mm.

11 mm. — 21 days	40 mm. — 56 days
17 mm. — 35 days	220 mm. — 100 days
28 mm. — 49 days	300 mm. — 115 days (full term)

Please bear in mind, as you dissect your fetal pig, that there are very few differences between the anatomy of the pig and that of the human being. As you dissect out each system, make a mental comparison between yourself and the fetal pig concerning the arrangement of the internal organs.

III.   External Anatomy

Beginning at the anterior end, locate the **mouth**, which leads into the **oral cavity**, the **nose**, the **nostrils**, which lead into the **nasal cavity**, the **ears**, the **external ear canal**, which leads inward from the ears. Locate the **nictitating membrane** in the corner of the eye, the **eyeball, eyelashes**, and **eyelids.**

Now, lay the pig on its side and identify the major body divisions beginning anteriorly with a large **head**, a short thick **neck**, a cylindrical **trunk** with two pairs of **appendages**, and a short **tail.** The **anus** is located ventral to the tail.

Examine the forelimbs and locate the **shoulder, elbow**, and **foot.** On the hindlimbs find the **hip, knee, hock joint** (ankle), and the **foot.**

Turn the pig ventral side up and locate the large **umbilical cord** in the abdominal region. This cord connects the fetus to the mother at the placenta. Cut a half inch off the umbilical cord and observe the umbilical blood vessels. The blood vessels in the umbilical cord consist of two small **umbilical arteries** having relatively thick walls and an **umbilical vein**, considerably larger than the two arteries. The umbilical arteries carry blood from the fetal pig to the placenta, while the single umbilical vein returns blood to the fetal pig's body from the placenta. A fourth very small vessel, the **allantoic duct**, serves to carry some of the small amount of urine formed by the kidneys away from the fetus. Refer to Figure 22.1. Make a sketch below of a section through the umbilical cord, and label the four vessels.

CROSS SECTION OF UMBILICAL CORD

Also, on the ventral surface on either side of the umbilical cord, find the row of **nipples** or **teats.** The number of nipples indicates the number of mammary glands.

Next, determine the sex of your pig. You will be expected to recognize both male and female pigs, even though you have only one sex for dissection. In the male, the opening of the urogenital tract, which serves both reproductive and excretory functions, lies at the end of the **penis**, just posterior to the umbilical cord. The **scrotum**, which contains the **testes**, is situated ventrally with respect to the anus. The female pig has a single urogenital opening, located ventral to the anus, with a dorsal projection of tissue, the **clitoris** or **genital papilla** below it.

IV. Dissection of the Fetal Pig

For this dissection you will need string, scissors, a sharp scalpel or a single edge razor blade, a blunt probe, forceps, and some dissecting pins. Place the pig on its dorsal surface in the dissecting pan. Referring to Figure 22.1, tie a string around one forelimb and place the string around and under the pan and tie it to the other forelimb. Make sure that the forelimbs are spread apart. Repeat this procedure for the hindlimbs. Note: When preparing to put the pig back in the storage container for future use and reference, do not untie the string. Simply slide the pan from underneath the pig. Also, put a tag on the pig, identifying the members of the group. Use a pencil when writing on the tag to avoid fading.

Dissecting does not mean cutting up the specimen. Instead, it means exposing a specimen to view. Use the scalpel or razor blade carefully and sparingly. You may find that the most useful tool is the dull probe, which can be used to separate organs from membranes. *Be very careful* with your dissection. Any tissues or organs that are removed should be deposited in a trash can and *never in the lab sink.*

A.   Respiratory System

This system functions in the exchange of gases between the internal and external environments of the organism. Air enters and leaves through the **nostrils** which lead, via the nasal cavity, into the **pharynx** or throat. The pharynx is where food and air passages cross and is located posterior to the **oral cavity.** Expose the pharynx by inserting scissors into the corners of the mouth and cutting the jawbones. You may have to cut approximately one to one and a half inches. Separate the jaws further by pushing down on the tongue until you find a cartilaginous projection, the **epiglottis** at the base of the tongue. This flap aids in preventing food from entering the air passageway which leads into the lungs. The epiglottis covers an opening known as the **glottis,** the opening of the windpipe or trachea.

Figure 22.1 indicates the incisions required to open the thoracic and abdominal regions. A scalpel and forceps should be used during this part of the dissection. Make cuts as indicated by the dotted lines in Figure 22.1. To enter the abdominal cavity, you will pass through an outer layer of skin, several layers of muscle, and a tough inner glistening membrane, the **peritoneum,** which lines the body cavity and surrounds the internal organs.

Pick up the umbilical cord and pull lightly on it. Passing from the base of the umbilical cord to the liver, the large brown organ covering most of the anterior portion of the abdominal cavity, is the **umbilical vein.** After identifying it, cut it, leaving two stump ends that can be located later.

A certain amount of brown liquid may be present in the body cavity. This is clotted blood and it should be poured into the sink; after which you should completely rinse the body cavity with cold tap water. Now, pin back the skin flaps, and fold back between the hind legs the skin and muscle to which the umbilical cord is attached.

Locate the muscular **diaphragm** which separates the thoracic cavity from the abdominal cavity, and aids in the movement of gases into and out of the lungs.

To expose the viscera in the thoracic region, continue the initial incision from the abdominal cavity forward to the clump of hairs under the chin. Gradually deepen the incision until you have cut through the **sternum,** or **breastbone.** The muscle is particularly thick in the neck region. Avoid cutting the organs and blood vessels in the neck and chest regions. Sever the edges of the diaphragm where it is attached to the rib cage. Next, break the rib cage by applying pressure from your thumb on the sternum. The flap of tissue which contains the sternum can be carefully trimmed away and discarded. Some of the ribs may have to be cut with scissors. If so, be careful not to cut into the organs of the thorax.

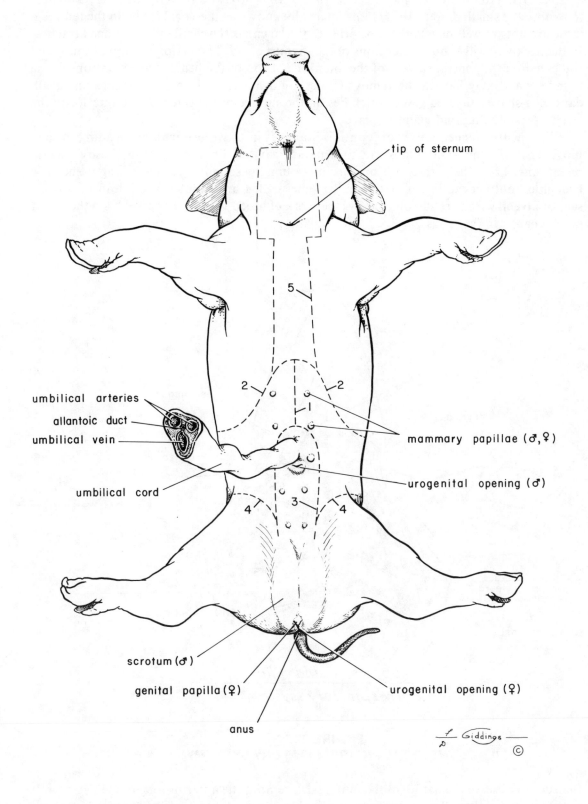

tip of sternum

5

2   2

umbilical arteries

allantoic duct

umbilical vein

mammary papillae (♂,♀)

umbilical cord

urogenital opening (♂)

4   3   4

scrotum (♂)

genital papilla (♀)

urogenital opening (♀)

anus

Giddings
©

FIGURE 22.1
THE FETAL PIG (EXTERNAL FEATURES AND DISSECTION GUIDELINES)

Referring to Figure 22.2 locate the **larynx,** or voicebox, and slit it open along its midline to expose the small, paired lateral flaps internally known as the **vocal folds.** In the fetal pig, these are not yet well-developed. Posterior to the larynx is the cartilaginous-ringed **trachea.** In the neck you will expose extensions of the **thymus gland.** The major portion of this gland is located on the ventral surface of the anterior portion of the heart. The extensions of the thymus gland may have to be removed. Also, in a midventral position, locate the small, dark, pea-shaped **thyroid gland** which lies on the surface of the trachea just anterior to the heart. Leave the thyroid gland in place.

Follow the trachea posteriorly until it branches into two **bronchi,** one leading to each **lung.** The left and right lungs are in separate cavities lined by a thin, fleshy **pleural membrane.** Trace the path of a bronchus into a lung by scraping away the lung tissue. The bronchus continues to branch into smaller tubes called **bronchioles** which lead to tiny air sacs or **alveoli** which are surrounded by a network of blood vessels. The alveolus is the site of gas exchange in the lungs.

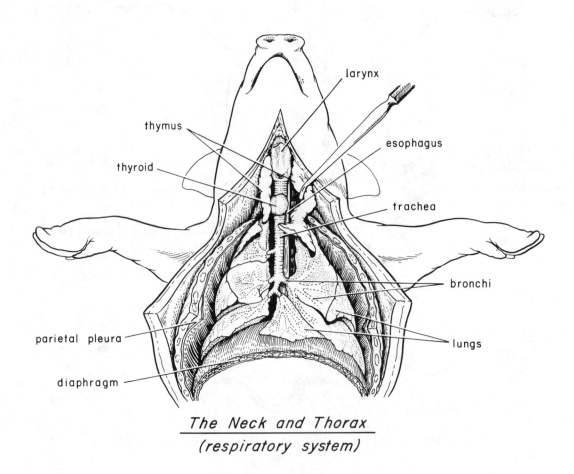

The Neck and Thorax
(respiratory system)

FIGURE 22.2
THE FETAL PIG (RESPIRATORY SYSTEM)

In order to observe the nature of the lung tissue, remove the anterior lobe of the left lung. It should be noted that all dissections regarding the pig refer to the body of the pig. Therefore, if the directions identify an organ on the left side, this refers to the left side of the pig. Place the lung in a small dish of water. Holding it with your forceps, gently tease the lung apart with the blunt wooden base of a probe. Using a stereomicroscope, identify the branching network of bronchioles and blood vessels.

Having examined the respiratory system in the pig, consult the human torso model and the wall chart of the human respiratory system. Identify the main structures of this system.

B.   Digestive System

Digestion begins in the **oral cavity** with the chewing of food. The **tongue** is attached at the back of the oral cavity. Observe the small, underdeveloped **teeth** in both the upper and lower jaws. The first set of teeth in mammals is called **milk teeth.** These are later replaced by the **permanent teeth.** Referring to Figure 22.3 note the ridged surface on the roof of the oral cavity. This is the **hard palate.** Posterior to the hard palate is the **soft palate.**

Locate the pharynx. Food leaves the pharynx via the **esophagus** on its way to the stomach. The opening into the esophagus is dorsal to the glottis. Insert a blunt probe into the opening and follow the esophagus posteriorly through the thoracic cavity. In the thorax, the esophagus lies just dorsal to the trachea. The esophagus penetrates the diaphragm and continues posteriorly to join the bag-like organ, the **stomach.** The **cardiac sphincter** is the muscle which forms the boundary between the esophagus and the stomach. Refer to Figure 22.3 in identifying the rest of the digestive system.

The stomach can be divided into the larger anterior **cardiac** portion and the lower tapering **pyloric** portion. This posterior segment of the stomach joins the small intestine. Open one side of the stomach and examine its interior surface. Does it appear smooth or rough? Locate the **liver,** the large lobed, reddish-brown organ which lies posterior to the diaphragm. Notice that the liver consists of several lobes which are attached only at the dorsal and anterior margins. Among other functions, the liver produces **bile,** which emulsifies fats in the small intestine. The bile is temporarily stored in a small sac, the **gall bladder.** This organ may be found by lifting up the extreme right lobe of the liver to which the gall bladder is attached. At the posterior end of the stomach, locate the hard ring of smooth muscle, the **pyloric sphincter.** This sphincter muscle forms the boundary between the stomach and the **small intestine.** Constriction of this muscle prevents food from escaping into the small intestine before the stomach has finished processing it.

The portion of the small intestine into which the stomach empties is the **duodenum.** Locate a long, whitish, cauliflower-like organ, the **pancreas,** lying dorsal to the duodenum and the stomach. The digestive enzymes produced by the pancreas pass into the duodenum. Try to locate the **pancreatic duct.** It may be difficult to find. Also locate the **bile duct** as it enters the duodenum from the gall bladder.

To the left of the stomach observe the **spleen,** a reddish brown tongue-shaped organ which functions in destroying old red blood cells in the adult. Notice that the small intestine is coiled, therefore providing an increased area for digestion and absorption of food. The small intestine is also held in place by a mass of sheetlike membranes, the **mesenteries.** Slit open a short portion of the small intestine and find the **villi** with a stereomicroscope. The villi are microscopic finger-like projections which serve to increase surface area also.

The small intestine continues posteriorly, merging with the first of three segments of the large intestine, the **colon,** which is a compact, rounded mass of intestine bound firmly by mesentery. Find the **caecum,** a blind pouch located posterior to the junction of the small and large intestines. The pig does not have an appendix. In man, the appendix is attached to the posterior end of the caecum. The straight most posterior section of the large intestine is the **rectum** which opens to the outside by way of the **anus.** The caecum, colon, and rectum make up the large intestine.

If time allows, sever the coiled small intestines just below the duodenum, and sever the colon at the point where it joins the rectum. Carefully cut the mesenteries holding the long intestinal sections in place so that it can be laid out in a straight line. How long are the intestines?

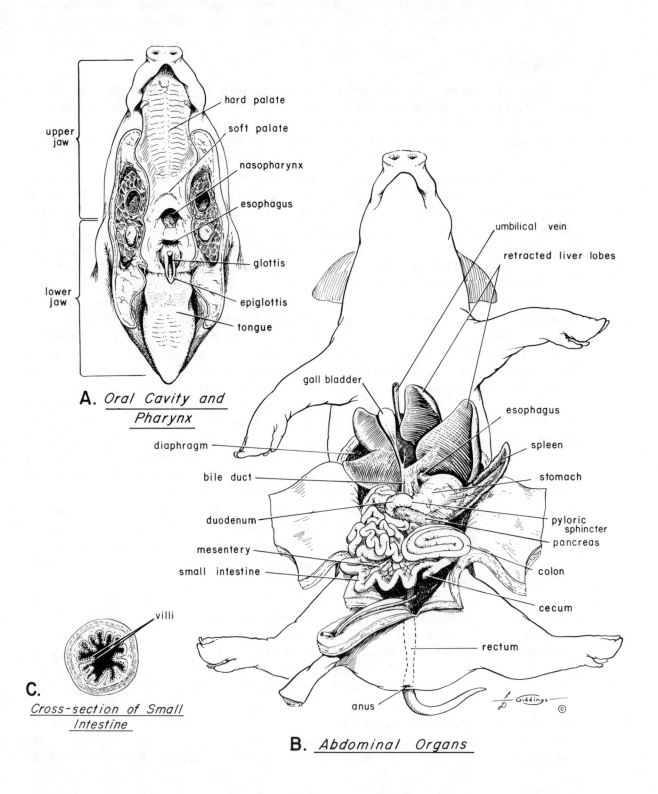

A. *Oral Cavity and Pharynx*

hard palate
soft palate
nasopharynx
esophagus
glottis
epiglottis
tongue

upper jaw
lower jaw

umbilical vein
retracted liver lobes

gall bladder

diaphragm
bile duct
duodenum
mesentery
small intestine

esophagus
spleen
stomach
pyloric sphincter
pancreas
colon
cecum
rectum
anus

villi

C.
Cross-section of Small Intestine

B. *Abdominal Organs*

FIGURE 22.3
THE FETAL PIG (DIGESTIVE SYSTEM)

Locate as many of the structures as possible in the digestive system on the human torso model and human anatomy wall chart.

C. Circulatory System

The major **arteries** and **veins** are injected with latex. The arteries, which carry blood away from the heart, are injected with red latex, and the veins, which carry blood toward the heart, are injected with blue. Sometimes the pressure of injection causes the latex to cross capillary beds in some places. Therefore, veins occasionally contain some red latex, while arteries may have blue latex.

1. Arterial System (Refer to Figure 22.4, 22.6 and 22.7)

Locate the **heart** in the thoracic cavity and carefully trim away the **pericardial sac** which surrounds the heart. The major portion of the heart is composed of the **right** and **left ventricles.** These are separated from one another on the ventral surface by the **coronary artery.** The left ventricle alone makes up the posterior tip, or **apex,** of the heart. Anteriorly and laterally are two dark projections, the right and left **auricles,** which are parts of the upper heart chambers, the **atria.** The term "auricle" comes from a root meaning "ear." The projections looked like ear flaps on each side of the heart. The auricles do not have any particular function.

Locate the **pulmonary trunk.** It is a large vessel which carries blood from the right ventricle of the heart. This large artery crosses from the upper right over the ventral surface of the heart and arches to the dorsal side of the heart, where it branches. Carefully remove the connective tissue surrounding the pulmonary trunk and trace it until it branches by displacing the heart to the right (the pig's right) side. The **right pulmonary artery** goes to the right lung and the **left pulmonary artery** goes to the left lung. Next, locate the large, thick-walled **aortic arch,** or **aorta,** which carries blood from the left ventricle. It passes anteriorly for a short distance and makes a sweeping 180° left turn and comes to lie posteriorly along the dorsal wall of the thoracic and abdominal cavities.

Observe that the pulmonary trunk connects directly to the aorta on the left side of the heart. After the trunk branches to the right and left lungs, the short, thick interconnecting vessel is known as the **ductus arteriosus.** The ductus arteriosus shunts blood into the **dorsal aorta,** bypassing the lungs. The lungs are nonfunctional in the fetal pig. After birth, the arterial duct normally closes off with connective tissue.

Carefully dissect away the connective and muscle tissues in the neck to expose the arterial branches going to the head.

Two major arterial branches emerge from the aortic arch a short distance from the heart. The first is the **innominate artery** or **brachiocephalic artery** which supplies blood to the right forelimb and head. This branch continues toward the head, then divides into two branches. The first is the **right subclavian artery** which supplies blood to the right forelimb and right ventral chest wall. The second branch continues anteriorly, dividing into two **common carotid arteries,** which supply blood to the head. The second branch off of the aortic arch is the **left subclavian artery** which carries blood to the left forelimb and the left ventral chest wall.

Find the dorsal aorta in the abdominal cavity. The first major branch is the **coeliac artery** which supplies the stomach, spleen and liver. The **anterior mesenteric artery** branches from the dorsal aorta just posterior to the coeliac artery. It supplies blood for the pancreas and the small intestine. Continue tracing the aorta posteriorly and locate the **renal arteries,** one leading to each kidney. Posterior to the renal arteries locate the paired **genital arteries** which are thread-like in size and lie on the extreme ventral surface

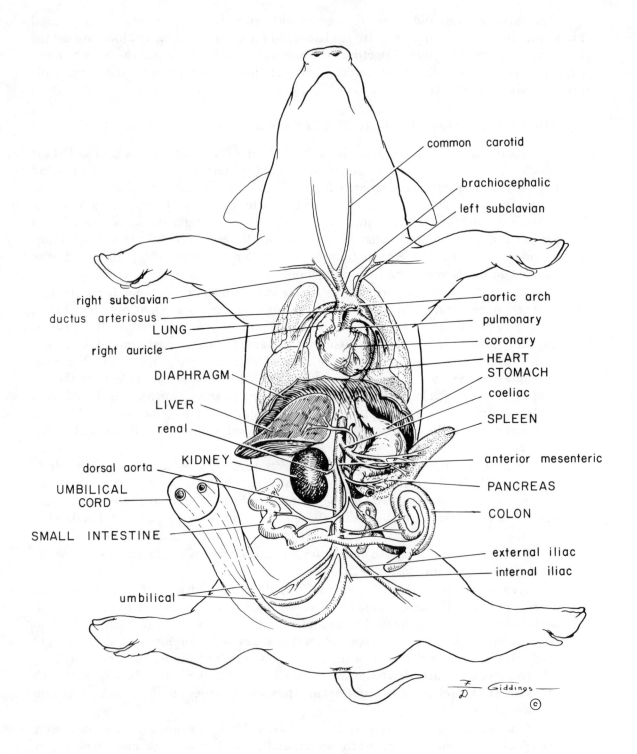

common carotid

brachiocephalic

left subclavian

right subclavian

ductus arteriosus

LUNG

right auricle

DIAPHRAGM

LIVER

renal

KIDNEY

dorsal aorta

UMBILICAL
CORD

SMALL INTESTINE

umbilical

aortic arch

pulmonary

coronary

HEART
STOMACH

coeliac

SPLEEN

anterior mesenteric

PANCREAS

COLON

external iliac

internal iliac

FIGURE 22.4
THE FETAL PIG — MAJOR ARTERIES

226

of the dorsal aorta. The genital arteries lead and supply blood to the sex organs. At the posterior end of the body cavity, the dorsal aorta divides into two pairs of arteries that supply blood to the legs, the **external iliac arteries** and the **internal iliac arteries.** The two large **umbilical arteries** pass ventrally from the internal iliac arteries.

2. Venous System (Refer to Figure 22.5)

Find the large vein which enters the right atrium anteriorly. This is the **anterior vena cava** or **precaval vein.** There are four major veins which drain the anterior region of the body that unite to form this blood vessel. The two **external jugular veins** lie parallel to the carotid arteries. These are paralleled by two **internal jugular veins.** Also locate the **left** and **right subclavian veins** which drain each of the forelimbs, respectively.

Locate the **azygos vein** which lies to the left of and parallel to the dorsal aorta in the chest cavity. The azygos vein drains the muscles between the ribs and empties directly into the right atrium.

Locate the large **posterior vena cava** or **postcaval vein** which runs parallel to the dorsal aorta in the lower abdomen. Note where the two **renal veins** (one from each kidney) join the postcava at the kidneys. Posteriorly, the posterior vena cava will divide into the paired **common iliac veins.** Further subdivision of the common iliac veins does occur, but the veins are difficult to find.

The organs of the gastrointestinal tract are drained by the **hepatic portal system.** Try to find the **ductus venosus** which may have been cut when you opened up the pig. Locate the **hepatic portal vein** carrying blood from the intestine to the liver. In the liver the blood passes through a capillary bed, where toxic materials are removed from the blood and the nutrient content of the blood is regulated by cells of the liver. Blood leaving this capillary bed eventually enters the posterior vena cava via the **hepatic veins.**

3. The Heart

Do not remove the pig's heart. For this dissection, you will use the sheep's heart.

Open the heart by making a midventral slit if this has not already been done. Use Figure 22.7 as a guide. Blood leaving the right atrium passes through a one-way valve, the **tricuspid valve,** into the right ventricle. Find the **bicuspid,** or **mitral valve,** in the same position in the left side of the heart. The tough cords holding the edges of the valves in place are called **chordae tendinae.** The chordae tendinae, which prevent the valves from flapping up into the atria when the ventricles contract, are attached posteriorly to **papillary muscles,** columns of muscle arising from the wall of the ventricle. In the fetal pig, the **foramen ovale,** an opening between the two atria, allows the lungs to be bypassed. It is reduced to a closed depression at birth. Locate this depression in the interatrial septum. Look down into the stub of the aortic arch and pulmonary trunk and observe the flaps that compose the **semilunar valves.** These valves prevent backflow of blood from the arteries.

After finding the bold face parts given previously on the sheep heart, locate the same parts on the beef heart on display and the human torso model.

D. Urinary System (Refer to Figure 22.8)

Locate the paired **kidneys,** lying dorsally against the abdominal wall. Notice that each kidney is covered by a thin membrane, the **peritoneum,** on its ventral surface. Next, find the **adrenal gland,** a narrow whitish body about half an inch long which lies medially along the anterior edge of the kidney. Remove the peritoneum from a kidney. On the medial side of the kidney is a concave depression, the **hilum.** At this point the **renal artery** and **renal vein** attach, carrying blood into and out of the kidney, respectively. The hilum is also the point where the **ureter** leaves the kidney, carrying urine to the **urinary bladder.** Expose one of the ureters by picking away the peritoneum covering it.

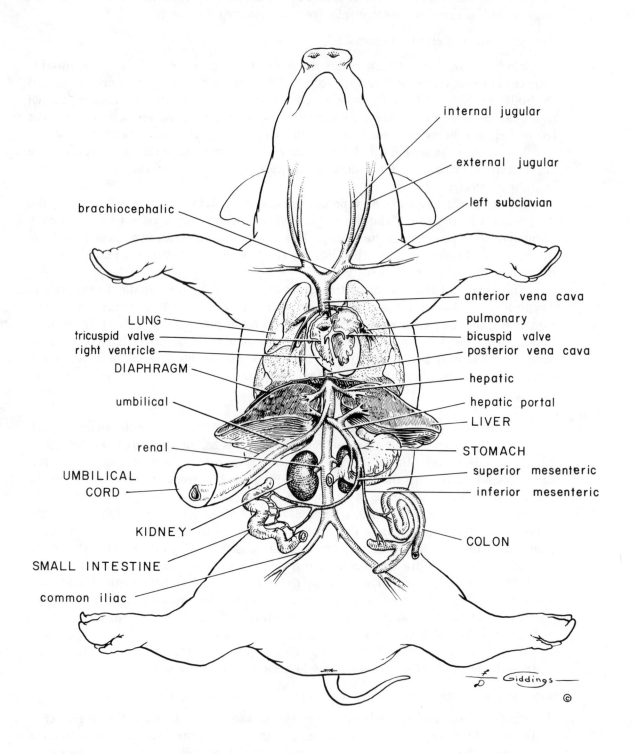

internal jugular

external jugular

left subclavian

brachiocephalic

anterior vena cava

LUNG

pulmonary

tricuspid valve

bicuspid valve

right ventricle

posterior vena cava

DIAPHRAGM

hepatic

umbilical

hepatic portal

LIVER

renal

STOMACH

UMBILICAL
CORD

superior mesenteric

inferior mesenteric

KIDNEY

COLON

SMALL INTESTINE

common iliac

Giddings

©

FIGURE 22.5
THE FETAL PIG **MAJOR VEINS**

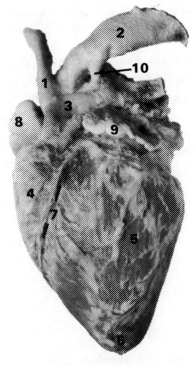

1. BRACHIOCEPHALIC ARTERY
2. AORTA
3. PULMONARY ARTERY
4. RIGHT VENTRICLE
5. LEFT VENTRICLE
6. APEX
7. ANTERIOR LONGITUDINAL SCILCUS
8. RIGHT AURICLE
9. LEFT AURICLE
10. DUCTUS ARTERIOSUS

FIGURE 22.6
SHEEP HEART LATERAL

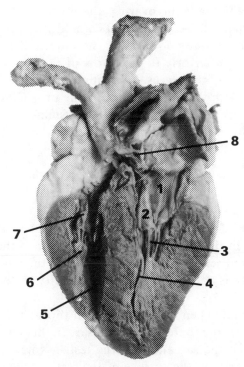

1. LEFT ATRIUM
2. BICUSPID VALVE
3. CHORDAE TENDINAE
4. LEFT VENTRICLE
5. RIGHT VENTRICLE
6. PAPILLARY MUSCLE
7. TRICUSPID VALVE
8. SEMILUNAR VALVE

FIGURE 22.7
SHEEP HEART (L.S.)

229

During fetal life, urine exits from the anterior end of the urinary bladder by way of the **allantoic duct,** through the umbilical cord to the placenta. In the adult male, urine is voided to the outside of the body by the **urethra.** This is a tube-like structure which passes posteriorly from the urinary bladder for a short distance and then turns anteriorly and ventrally to enter the **penis.** In the adult female the urethra continues posteriorly to enter the **urogenital sinus,** or **vestibule,** which lies approximately one-half of an inch from the urogenital opening.

Using your razor blade or scalpel, section one of the kidneys in place, cutting from the lateral or the medial side on a plans parallel to the dorsal side of the animal. Note that at the center of the medial portion of the kidney is an irregular cavity, the **pelvis.** Here the urine is collected and through this region the branching blood vessels pass. The pelvis is continuous with the ureter.

The outermost portion of the kidney, the **cortex,** shows many small striations perpendicular to the outer surface. This region and the **medulla** region, which lies medially to it, are composed of thousands of minute excretory tubules called **nephrons** associated intimately with capillaries. Urine is formed in these regions and then drains to the pelvis.

E.   Reproductive System

The reproductive system in mammals is closely associated with the urinary system. Together, these form the **urogenital system.** In this exercise, they will be discussed separately, but in Figure 22.8 they are combined.

1.   Female Reproductive System

Pull the umbilical cord posteriorly and find the urinary bladder. The urinary bladder continues posteriorly joining the urogenital sinus via the urethra. Cut along one side of the urogenital sinus and through the cartilage of the pelvic girdle. Now you should be able to lay the legs out flat.

Referring to Figure 22.8, locate the pair of gonads the **ovaries,** about one half an inch posterior to each kidney. The ovary is a small, bean-shaped, light colored structure. The lateral and dorsal surface of each ovary is partially covered by the mouth of the **oviducts,** or **fallopian tubes.** The oviducts lead posteriorly to the **horns of the uterus.** The horns join along the midline to form the **body of the uterus.** Posterior to the body of the uterus, locate the **vagina.** The vagina joins with the **urethra** forming the **urogenital sinus,** or **vestibule,** an area shared by the urinary and reproductive systems. The urogenital sinus opens to the outside via the **urogenital opening,** just ventral to the anus. Now locate the dorsally directed **clitoris,** or genital papilla, the female homolog to the penis in the male.

2.   Male Reproductive System

Referring to Figure 22.8, locate the pair of gonads, the **testes,** where sperm is produced. The testes begin their development in the body cavity, just posterior to the kidneys. Before birth they descend into the paired **scrotal sacs,** located between the hind legs and just ventral to the tail. Each scrotal sac is connected to the body cavity by the **inguinal canal.**

Locate the opening of the left inguinal canal. Then make an incision through the skin and muscle layers from a point over this opening to the left scrotal sac. In this way, the canal and whole sac will be exposed. Open the sac and find the testis. Note the much-coiled tubule, the **epididymis,** which lies along the surface of the testis. This is continuous with the sperm duct, or **vas deferens** which passes back toward the body cavity. In the body cavity, the sperm ducts from each testis loop over the umbilical

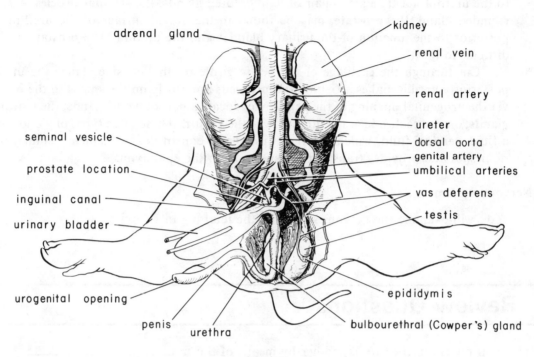

adrenal gland

kidney

renal vein

renal artery

ureter

dorsal aorta

genital artery

umbilical arteries

seminal vesicle

prostate location

inguinal canal

urinary bladder

vas deferens

testis

urogenital opening

epididymis

penis

urethra

bulbourethral (Cowper's) gland

Male

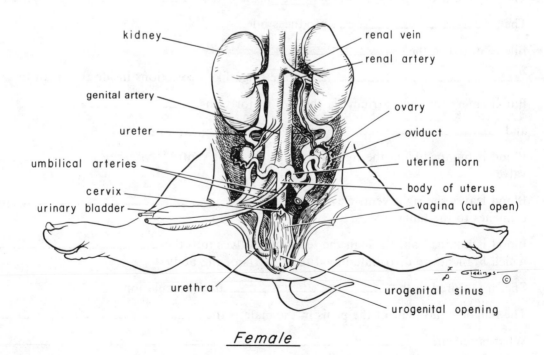

kidney

renal vein

renal artery

genital artery

ovary

ureter

oviduct

umbilical arteries

uterine horn

cervix

body of uterus

urinary bladder

vagina (cut open)

urethra

urogenital sinus

urogenital opening

Female

FIGURE 22.8
THE FETAL PIG (UROGENITAL ORGANS)

231

arteries and ureters and unite dorsally at the posterior end of the urinary bladder with the **urethra.**

Near where the vasa deferentia (singular: vas deferens) join the urethra and dorsal to the urethra, locate a small pair of light colored glands, the **seminal vesicles.** A small rounded gland, the **prostate,** may be found in the dorsal surface of the urethra just posterior to the junction of the urinary bladder and the urethra. The prostate may be difficult to locate.

Cut through the cartilage of the pelvic girdle on the left side. Trace the urethra posteriorly until it makes a U-turn and proceeds anteriorly until it opens to the outside via the urogenital opening of the penis. Now locate a pair of white glands, the **Cowper's glands,** one on each side of the urethra near the U-turn. These three types of glands form a fluid, **seminal fluid,** which together with sperm constitutes the **semen.** Seminal fluid nourishes and provides transport for sperm from the epididymis.

F.   Nervous System

You will examine the nervous system of the fetal pig in Exercise 24.

V.   Review Questions

A.   The fetal pig is attached to its mother by means of the _____ cord.

B.   The sites of gas exchange in the lungs are tiny sacs called _____ .

C.   The flaplike projection which covers the glottis is the _____ .

D.   The _____ produces bile.

E.   Bile is stored in the _____ .

F.   _____ are fingerlike projections inside the small intestine.

G.   Blood enters the right atrium via the two major veins, _____

and _____ .

H.   Blood is pumped from the right atrium to the right ventricle via the _____ valve.

I.   Blood leaves the right ventricle via the _____ which branches to the lungs.

J.   Blood is pumped out through the semilunar valves into the _____ which branches to distribute blood to all parts of the body.

K.   Sperm is stored in the _____ until ejaculation.

L.   The female equivalent of the penis in the male is the _____ .

M.   What is semen? _____

N.   Describe the function of seminal fluid. _____

_____

O.   The thin membrane covering the kidneys is called the _____ .

P.   Describe the function of the placenta. _____

_____

Q.   How old is your pig? _____

R.   Are mammary glands present in both sexes? _____ How many? _____

S.   How many lobes do you find for each lung? _____

T.   Does the stomach interior appear smooth or rough? _____

Any advantage to this? _____

U.   Define emulsify. _____

_____

V.   What is the function of the foramen ovale? _____

_____

W.   What happens to this opening at birth? _____

CIRCULATION

I. Objectives

After the completion of this exercise, the student should be able to do each of the following:

A. List the main components of blood.

B. Determine by microscopic examination whether blood is that of a mammal or non-mammal.

C. Upon microscopic examination, distinguish among each of the following: **erythrocyte, neutrophil, eosinophil, basophil, lymphocyte,** and **monocyte.**

D. Interpret results from a simple blood-typing test.

E. Determine clotting time of blood.

F. Explain the relationship between hemoglobin levels in blood and the colors observed on the Tallquist chart.

G. Determine the hematocrit of the blood and explain what is indicated by the results.

H. Identify the origins of the heart sounds heard through a stethoscope.

I. Determine heart rate and pulse rate and tell how they vary with exercise.

J. Use a sphygmomanometer to determine blood pressure and explain what the figures stand for.

K. Locate a valve in an arm vein.

L. Answer the questions throughout this exercise including the review questions at the end.

II. Introduction

In small organisms consisting of one or a relatively few cells, oxygen and dissolved food can diffuse directly into the protoplasm of the cell from the environment. Carbon dioxide and other wastes produced in the cell can diffuse into the environment. In larger, multicellular organisms, however, most cells are too far from the external environment for simple diffusion of nutrients and wastes to be adequate. In higher animals, well-developed circulatory systems transport materials throughout the body. Blood, the circulatory fluid of higher animals, is distributed throughout the body via a network of vessels (arteries, capillaries, and veins). The primary force behind its movement is the contraction of the heart.

Blood is composed of two basic parts: a fluid portion called **plasma,** and solid matter suspended in the plasma. The suspended matter consists of small cytoplasmic fragments called blood **platelets** and different kinds of **corpuscles,** or blood cells; these include the red blood corpuscles, or **erythrocytes,** and the white blood corpuscles, or **leukocytes.**

The erythrocytes, which are much more numerous than the leukocytes, transport oxygen. They contain a red pigment called **hemoglobin,** which combines readily with oxygen in a loose compound called **oxyhemoglobin.** Oxyhemoglobin is a brighter red than hemoglobin; this is why blood from an artery, which contains oxyhemoglobin, is a brighter red than blood from a vein, which contains oxygen-poor blood. The red blood cells are rounded disk-shaped cells without nuclei when mature. They do not live very long and are constantly replaced.

Leukocytes are usually larger than erythrocytes, and each one has a nucleus. They move somewhat like and, hence, are said to have amoeboid motion. White blood cells as a group have several functions, one of which is to engulf bacteria and other foreign organisms which might invade the body. Certain of these cells are attracted to points of infection, where they accumulate and do battle with the invading bacteria. In the process, some of the white cells are killed and some of the surrounding tissue is broken down. The combination of bacteria, broken-down tissue, and white blood cells forms the whitish semiliquid called **pus** that sometimes accumulates around a wound.

Blood platelets, cell fragments, are very small bodies that are important in the clotting of the blood.

This laboratory exercise may be said to have two purposes: first, to study blood cells and, second, to introduce certain laboratory techniques used in the study of blood and circulation.

III. Comparison of Mammalian and Non-Mammalian Blood

Obtain slides of blood smears from various types of organisms and make sketches of the blood cells seen in the space provided. Identify and label the red blood cells (RBC's) and white blood cells (WBC's) in each slide. The red blood cells will always be the most numerous cells.

TURTLE BLOOD   **FROG BLOOD**

CHICKEN BLOOD   **RAT BLOOD**

What characteristic can be used to distinguish between mammalian and non-mammalian blood?

What selective advantage could this characteristic confer on mammals?

How does the presence or absence of a nucleus in a RBC affect (1) its life span, (2) its size, (3) its efficiency in transporting oxygen?

IV.   Morphological Characteristics of Blood

Stained blood smears are customarily used to examine and identify the WBC's. Examine a stained smear of human blood under low power and high power of the microscope. Refer to Figure 23.1 and to colored figures in a reference text.

A.   Erythrocytes (RBC's)

These are the most numerous objects seen in the stained smear. They are round, biconcave disks. The centers of the **erythrocytes** may appear light because the cells are thinnest there.

B.   Leukocytes (WBC's)

Although there are a number of ways of classifying the **leukocytes** of man, the following outline of names and descriptions is generally accepted:

1.   Granular Leukocytes (Granulocytes)

These are of several kinds, all of which have characteristically lobed nuclei and all of which have granular cytoplasm. Because of the lobed nature of the nuclei these cells are sometimes designated as **polymorphonucleocytes,** or "polymorphs". They are produced mainly in the red bone marrow. When granular leukocytes are stained with certain dyes, such as the mixture known as Wright's stain, the cytoplasmic granules do not all become colored the same way. On the basis of this differential staining, the granular leukocytes are further separated into groups, of which the following three kinds are usually listed:

237

a. **Neutrophil:** This kind of granular leukocyte has many very small rose-violet or purple granules distributed throughout the cytoplasm and is more abundant than all the other leukocytes combined. These comprise 54-62% of all leukocytes.

b. **Eosinophil:** This leukocyte has large distinctly red granules scattered throughout the cytoplasm. The nucleus most often shows only two to four lobes. Only 1-3% of the leukocytes are eosinophils.

c. **Basophil:** This kind of granular leukocyte has relatively few large dark blue granules distributed somewhat irregularly throughout the cytoplasm. A few granules may overlie the nucleus which is likely to be less distinctly lobed than that of other granular leukocytes. Often the nucleus of the basophil seems only slightly indented or irregular in outline. Basophils comprise less than 1% of all leukocytes.

2. Lymphocytes

These are the second most numerous of the leukocytes. The cytoplasm of these leukocytes may or may not contain small, either red or clear blue, granules, or a mixture

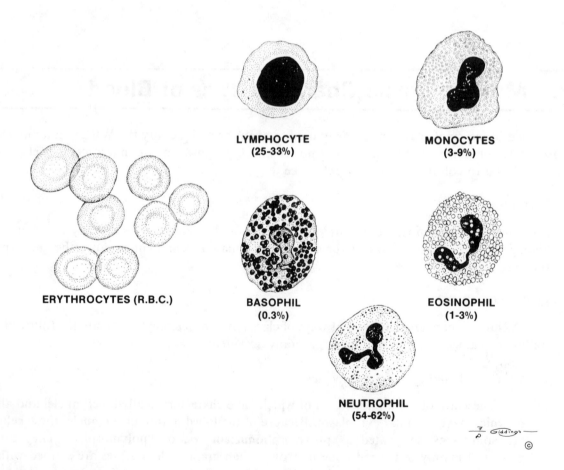

LYMPHOCYTE
(25-33%)

MONOCYTES
(3-9%)

ERYTHROCYTES (R.B.C.)

BASOPHIL
(0.3%)

EOSINOPHIL
(1-3%)

NEUTROPHIL
(54-62%)

FIGURE 23.1
ERTHROCYTES AND LEUKOCYTES

of the two. Generally, however, **lymphocytes** are recognized by the pale to dark blue cytoplasm which surrounds a round, oval, or slightly indented purple nucleus. Lymphocytes vary considerably in size. Smaller ones will have only a narrow band of cytoplasm around the nucleus, while larger lymphocytes will have a broader band of cytoplasm. Small lymphocytes have a diameter only slightly wider than that of an RBC. Large lymphocytes may have a size that is close to that of granulocytes. They are produced in lymphoid tissue such as that in the spleen and lymph nodes. Between 25-33% of all leukocytes are lymphocytes.

3. Monocytes

These cells are the largest of the leukocytes. They have a wide band of gray-blue cytoplasm containing very small clear blue granules which are generally apparent only under high magnification in a well-prepared and perfectly stained blood smear. The purple nucleus may be round, irregular in outline, or indented, but is so often "horseshoe" shaped that this characteristic is likely to be the best means of identifying them. Like the granular leukocytes, they are produced mainly in red bone marrow. From 3-9% of the leukocytes are **monocytes.**

C. Blood Platelets (Thrombocytes)

Find a cluster of blood **platelets.** These are not cells, but are particles of protoplasm separated from the cytoplasm of special giant cells in the bone marrow. They appear as tiny blue bodies, usually found in groups on stained smears.
What is the function of the platelets?

V. Physiological Properties of Blood

A. Blood Type

There are four basic types: A, B, AB, and O. Blood type is determined by the presence of certain proteins on the RBC; these proteins are known as **antigens.** In addition to protein factors on the RBC's, blood also contains certain proteins known as **antibodies** in the plasma. Some antibodies fight organisms entering the body. The disease-fighting antibodies usually develop after birth, but the blood antibodies are present at birth. A person never has the same type antibody as he does antigen. Why?

Blood Type	Antigen Type	Antibody Type	% Distribution (USA)
A	A	B	40
B	B	A	10
AB	A & B	none	5
O	none	A & B	45

In addition to the antigens mentioned above, a person may have still another blood antigen on his RBC's, the **Rh antigen.** If the antigen is present, the individual is considered Rh⁺. If the antigen is absent, the individual is Rh⁻. The Rh factor is of particular importance in the case of an Rh⁻ woman being pregnant with an Rh⁺ child. If, for some reason, blood leakage occurs between the child and the mother and the Rh⁺ blood comes into contact with the Rh⁻ blood of the mother, she may develop antibodies against the Rh antigen of the child. These antibodies can enter the second child's bloodstream via the placenta and cause clumping of the RBC's, leading to anemia, jaundice, or even death.

The importance of blood typing is best demonstrated in the case of whole blood transfusions (i.e., both cells and plasma are given to another individual). Both the donor and recipient must be blood typed to insure compatibility of their blood. If they are not compatible, clumping of RBC's (**agglutination**) will occur and may cause death. If a person with blood type A were to donate blood to another person with blood type B, what type of results would you expect? In this case, the recipient's A antibodies will cause a clumping of the RBC's (containing A antigens) he receives from the donor. These clumps can then cause blockage of tiny capillaries in the heart, brain, kidney, etc.

What blood types are compatible?

Would blood typing be critical if a person were to receive a transfusion of blood plasma only? Why?

The determination of a person's blood type is based on the antigen-antibody interaction causing agglutination. You will be using 3 solutions: Anti-A serum (blue, containing A antibodies), Anti-B serum (yellow, containing B antibodies), and Rh (Anti-D) serum, (clear, containing antibodies against Rh factor). You will mix a drop of your blood with each of these sera and, if clumping occurs, that particular antigen is present on your RBC's.

Procedure:

1. Draw 3 separate circles on a clean, dry slide with a wax marking pencil. Label the left as "A", the middle as "B", and the right "Rh".

2. Draw a drop of blood and touch it in the center of each circle.

3. Add a drop of Anti-A serum to the circle on the left, a drop of Anti-B to the middle circle, and a drop of Anti-D or Rh to the right.

4. Mix each, using a separate toothpick for each drop.

5. Let the mixtures stand for 2-3 minutes, gently agitating the slide to speed up the reaction (you may have to hold the slide over a light so that the heat will speed up the reaction).

6. Agglutination or positive reactions will be readily identified by the appearance of granules consisting of large clumps of RBC's. Be sure to make your observations before the slide dries out.

7. Have your instructor check your observations.

Conclusions:

B. Clotting Time of Blood

Rapid blood clotting is important in preventing excessive blood loss. This process is dependent on many different factors. Essentially what happens is that platelets, upon contact with collagen fibers exposed by the injury, release **thromboplastin,** which combines with calcium to convert **prothrombin** in the plasma to an active enzyme, **thrombin.** Thrombin acts upon another blood protein, **fibrinogen,** converting it to strands of **fibrin.** These strands form a mesh which traps cells and forms the framework for the development of a clot.

Procedure:

1. Draw a drop of blood, wiping away the first drop.

2. Hold the finger downward so that a large drop is suspended.

3. Place the drop inside a small piece of aluminum foil and fold it tight.

4. Wait one minute; then slowly unfold the piece of foil to a 45 degree angle. If the blood has started to clot, one or more thin threads of fibrin will be seen holding both sides of the foil together.

5. If fibrin fibers have not appeared, wait 30 seconds and repeat. Continue until fibers appear and note your clotting time. (Normal 2-6 minutes).

Conclusions:

C. Hemoglobin Level in the Blood

There are many different types of anemia, the most common of which is characterized by a deficiency of hemoglobin in the RBC's. Because of this deficiency, the oxygen-carrying capacity of the blood is lowered. An easy, but not extremely accurate, test for hemoglobin level is the Tallquist method, in which the color of fresh blood is compared with standard colors on a printed chart. The darker red the blood, the higher the hemoglobin concentration.

Below are the accepted hemoglobin percentages for normal and anemic blood.

Normal	*Anemia*
above 85% — male	below 70% — male
above 80% — female	below 70% — female

Procedure:

1. With a large drop of blood suspended from the finger, bring a piece of Tallquist test paper up to the drop, allowing the paper to absorb the blood slowly.

2. As soon as the blood loses its glossiness (but *before* it dries), place the smear below the holes in the scale and try to match it to the color closest to it on the chart.

3. Read your hemoglobin content from the chart.

Conclusions:

D. Hematocrit

The percentage of blood volume occupied by RBC's is known as the **hematocrit** or packed cell volume. It is used as a preliminary diagnostic test for anemias. If a tube of blood is properly centrifuged, the RBC's will pack into the bottom portion of the tube and the plasma will occupy the upper portion of the tube. From such a tube, it is easy to determine the percentage of volume occupied by RBC's.

Normal Values: Male — 40-54%   Female — 37-47%

Procedure:

1. Draw a drop of blood.

2. Using a heparinized capillary tube (red tip), fill the tube about ¾ full of blood. Heparin prevents the blood from clotting.

3. Seal off the red tip of the tube by pushing it into "critoseal".

4. Place the tube — red tip down — into a centrifuge tube with a bit of cotton in its bottom to cushion the capillary tube.

5. Place in centrifuge and spin at 2000 rpm for 5 minutes.

6. Remove capillary tube and, using a millimeter ruler, measure:

Height of total column _____

Height of cells only   _____   Hematocrit % _____

$$\text{Hematocrit \%} = \frac{\text{Height of cells only}}{\text{Height of total column}} \times 100$$

Conclusions:

VI.   Circulatory Physiology

Heart rate, pulse rate, and blood pressure can be used to indicate whether the heart and blood vessels are functioning normally. Figure 23.2 shows the flow of blood through the human heart.

A.   Heart Sounds

Beating of the heart is associated with a number of sounds, two of which can be heard readily with an ordinary stethoscope. The first sound is the result of the closing of the AV valves, caused by the contraction of the ventricles. It is a low-pitched "DUB" sound. The second is the result of the closing of the semilunar valves, due to the relaxation of the ventricles. When this occurs, blood falls back into the cuplike semilunar valves and causes them to "inflate" with a snap — like a parachute catching air. This produces a high-pitched "LUB" sound.

Procedure:

1.   To hear the first sound, place the stethoscope near the 5th or 6th rib, left of the center.

2.   To hear the second sound, place the stethoscope near the 2nd rib, slightly to the left of center.

When one of these sounds is not distinct, but is more of a gurgle, a person is said to have a

heart murmur. What could cause this? _____

_____

B.   Heart Rate

The speed at which the heart contracts is determined by many different factors, e.g., sex, age, size, physical condition, and mental state. The two heart sounds (LUB and DUB) make one complete heartbeat. The normal resting rate is 70-100 beats/min.

Procedure:

1.   Listen for the heart beat.

2.   Count the number of beats per 15 seconds. Multiply by 4 to obtain heart rate per

minute. _____

3.   Determine heart rate for each of the following conditions:

At rest (lying) _____

At rest (sitting) _____

At rest (standing) _____

After exercise (standing) _____

Conclusions:

C. Pulse Rate

One complete heartbeat corresponds to the pulse beat. The pulse beat is felt during the contraction of the ventricles. The most common place for the pulse to be taken is at the wrist. Average rate is 72 beats per minute (same as average heart beat). Why? Pulse rate is also affected by such things as age, sex, activity, and body temperature.

Procedure:

1. Place the fingertips gently but firmly over the radial artery along the inside of the wrist, below the base of the thumb.

2. Take a reading for 15 seconds, multiplying by 4 to obtain the pulse rate per minute.

_____

3. Determine the pulse rate for each of the following:

At rest (lying) _____

At rest (sitting) _____

At rest (standing) _____

After exercise (standing) _____

How long does it take for your pulse rate to get back to the value recorded at rest standing?

_____ (If it takes about 3 minutes or less, this indicates you are in good physical condition.)

Conclusions:

D. Blood Pressure

A blood pressure reading is obtained by listening for the pulse sounds in the brachial artery. As the blood is pumped into the blood vessels by the heart, it causes pressure to be exerted against the inner walls of the arteries. During heart contraction **(systole),** the blood pressure in the arteries reaches its maximum. During heart relaxation **(diastole),** blood pressure reaches its minimum. This blood pressure is measured indirectly by the use of a **sphygmomanometer.** Factors having greatest influence on blood pressure are force of heartbeat, resistance of artery walls, elasticity of artery walls, volume of blood, and viscosity of blood.

Procedure:

1. Place the cuff of the sphygmomanometer around the upper arm just above the elbow.

2. Shut off the valve near the rubber bulb and pump air into the rubber cuff by squeezing the bulb until the pressure reaches about 150 mm of mercury. This will stop the flow of blood through the brachial artery.

3. Place the stethoscope over the artery at the bend of the elbow; slowly release the thumb screw valve allowing air to leave the cuff.

4. Keep listening for blood seepage through the brachial artery. When the pressure in the cuff equals the pressure in the artery, the pulse will be heard distinctly. This thumping sound is produced by the contraction of the ventricles forcing blood into the arteries.

    This is the systolic pressure. _____ (Normally this varies from 90-140 mm.)

5. Continue to listen as the pressure in the cuff decreases. The pulse will eventually become

    very soft and disappear. Record the pressure when the pulse sound disappears. _____
    This is the diastolic pressure. (Normally this ranges from 60-90 mm.)

6. Measure blood pressure after exercising.

Express blood pressure as $\dfrac{\text{systolic}}{\text{diastolic}}$

Conclusions:

E. Valves in Veins

    The blood vessels taking blood back to the heart contain blood at a very low pressure. Therefore, to prevent backflow of blood, valves exist in veins. The location of these valves can be demonstrated as follows:

Procedure:

1. Expose right forearm. Compress the vessels near the elbow until the veins in the forearm stand out prominently.

2. Lay index finger on the left hand on one of the veins near the wrist; with thumb press the blood in the vein from the finger toward the heart.

3. If the blood flows back to the finger, lay the finger at the point the thumb reaches and repeat the process. Continue until a point is found beyond which the blood will not return to the finger. This indicates the location of a valve.

4. With the finger just below the valve, attempt, by pushing downward over the vessel with the thumb, to force the blood past the valve.

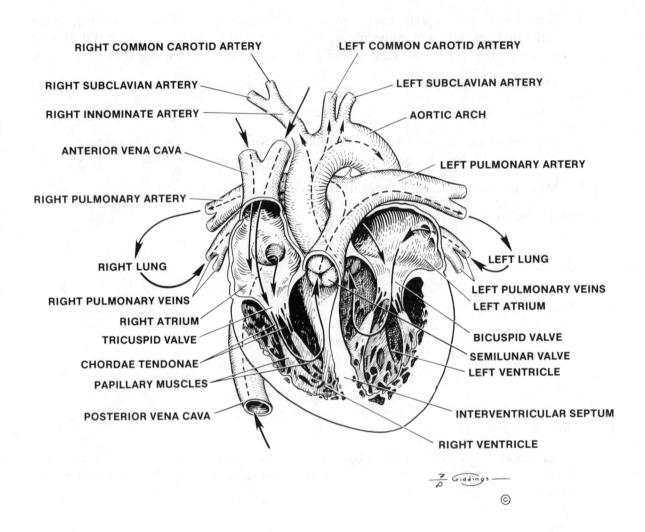

RIGHT COMMON CAROTID ARTERY

LEFT COMMON CAROTID ARTERY

RIGHT SUBCLAVIAN ARTERY

LEFT SUBCLAVIAN ARTERY

RIGHT INNOMINATE ARTERY

AORTIC ARCH

ANTERIOR VENA CAVA

LEFT PULMONARY ARTERY

RIGHT PULMONARY ARTERY

RIGHT LUNG

LEFT LUNG

RIGHT PULMONARY VEINS

LEFT PULMONARY VEINS

RIGHT ATRIUM

LEFT ATRIUM

TRICUSPID VALVE

BICUSPID VALVE

CHORDAE TENDONAE

SEMILUNAR VALVE

PAPILLARY MUSCLES

LEFT VENTRICLE

POSTERIOR VENA CAVA

INTERVENTRICULAR SEPTUM

RIGHT VENTRICLE

FIGURE 23.2
HUMAN HEART

Conclusions:

V. Circulatory Pattern in Frog's Foot

A live frog under anesthesia or with its brain removed has been positioned so that you can observe the web of the foot under low magnification. The web of thin skin between the toes affords an opportunity to observe the movement of RBC's through a capillary network. Note the rate of movement, the size of different vessels, the relative size of the RBC's and the distribution of the vessels. The black, irregular or star-shaped bodies are melanophores (pigment cells).

Does the direction of flow within a vessel reverse itself or does it always flow in one

direction? _____ Is the rate of flow correlated with size of vessel? _____
Is it possible for blood in these vessels to exchange gases with the external environment at this

point? _____

VI. Review Questions

A. What is indicated by clotting of the blood in each of the following sera:

    (1) Anti-A _____

    (2) Anti-B _____

    (3) Anti-D (i.e., Anti-Rh) _____

B. Name two ways of testing the blood for anemia. _____

_____

C. Why are the nuclei of the human white blood cells blue in a prepared blood slide?

_____

D. Why do the very numerous human red blood cells appear red? _____

_____

E. Why do the frog red blood cells appear red? _____

_____

F. Would a hematocrit of 30% be normal, or abnormal? _____

G. A person with low hemoglobin might be suffering from _____ .

H. Platelets are important in the _____ process.

I. Which type of blood agglutinates only with anti-Rh?

_____

J. State the function of white blood cells.

_____

K. Is a neutrophil an erythrocyte or leukocyte?

_____

THE NERVOUS SYSTEM

exercise 24

I. Objectives

After the completion of this exercise, the student should be able to do each of the following:

A. Name the twelve pairs of **cranial nerves** and give the function and origin of each.

B. Identify the major regions of the sheep, pig, and human brains, and know the major function of each.

C. Diagram, label, and explain the processes involved in a **reflex arc**.

D. Identify nervous tissue and its major components under the microscope.

E. Answer the review questions at the end of this exercise.

II. Introduction to the Nervous System

The nervous system of any animal, and especially that of man, is one of the most complex systems of the body from the standpoints of both structure and function. A complete understanding of the human nervous system is beyond the scope of this course, but certain fundamental concepts, as outlined in the objectives, should be understood.

The nervous system functions to integrate the activities of the body by interpreting and responding to internal and external stimuli. Then, in response to the commands of the nervous system, continuous adjustments are made by practically all tissues of the body, especially the muscles and the glands.

III. Nerve Tissue

The nervous system is composed of nerve tissue made up of nerve cells (**neurons**) and supportive tissue. In the brain and spinal cord, the supportive tissue is comprised of **glial cells**. In the cranial and spinal nerves, the supportive tissue is connective tissue.

Neurons are specialized for the reception of stimuli and the transmission, interpretation, and coordination of nervous impulses within the nervous system. Refer to your textbook for a diagram of the neuron.

Try to find a somewhat isolated neuron on a prepared slide. The **nucleus** is located in an enlarged portion of the cell called the **cell body**. Extending from the cell body are the nerve fibers which account for most of the cell length. Each vertebrate nerve fiber carries impulses in a specified direction. A **dendrite** picks up impulses and carries them toward the cell body and an **axon** carries impulses away from the cell body. It is not possible to distinguish dendrites from axons on your slide. Sketch a neuron from your slide and label the nucleus, the cell body, and a nerve fiber.

There are three functional types of neurons. A **sensory neuron** carries impulses toward while a **motor neuron** carries impulses away from the brain or spinal cord. **Association neurons** (or **interneurons**) are located within the brain or spinal cord.

IV.   Organization of the Nervous System

All of the neurons and their supportive cells collectively make up the nervous system which can be divided into the central and peripheral parts. The **central nervous system** includes the **brain** and **spinal cord** and the **peripheral nervous system** includes the **cranial nerves** leading to and from the brain, and the **spinal nerves** leading to and from the spinal cord. Nerves, in general, are bundles of nerve fibers (axons, dendrites, or both) with the neuron cell bodies being located in the central nervous system or as "bumps" (**ganglia**) along the nerves.

V.   Brain and Cranial Nerves

Observe the brain of the dissected fetal pig on demonstration in the lab. The tough outer covering of the brain (especially around the cerebellum) is called the **dura mater**. The brain is composed of three primary parts; (1) the **cerebrum (cerebral hemispheres)** continuing forward as the **olfactory lobes;** (2) the **cerebellum** posterior to the cerebral hemispheres and considerably smaller; and (3) the **medulla oblongata,** the enlarged anterior end of the spinal cord under the cerebellum (not easily seen from the view that you now have). Note the convolutions in the cerebrum — the folds are called **gyri** (singular — **gyrus**) and the depressions or grooves are called **sulci** (singular — **sulcus**).

Also available for study in the lab are preserved sheep brains. Use the labelled pictures in Figures 24.1-24.3 to find the parts of the brain. The cerebrum, cerebellum, and medulla are similar in appearance to the pig's brain. The two cerebral hemispheres are separated superiorly by the **longitudinal cerebral fissure** but connected inferiorly by the **corpus callosum** composed of nerve fibers connecting the two hemispheres. In a sagittal section of the sheep brain, the **pons, thalamus, hypothalamus, cerebral peduncles,** and the **corpora quadragemina** can be located. This view of the brain also shows that the interior of the brain has a number of **ventricles** (hollow chambers). Although the first and second ventricles are in the cerebral hemispheres and are not visible here, the third ventricle is housed within the thalamus, and the fourth ventricle is in the area of the medulla. On the underside of the brain, one can see the **olfactory lobes** (smaller than in the pig) and the structures associated with the entry of the optic nerve into the brain (**optic nerve, optic chiasma,** and **optic tract**).

Using Figure 24.4 as a guide, locate the brain structures on the human brain model.

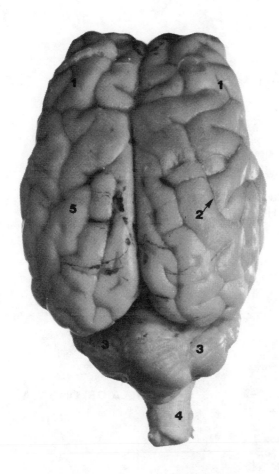

1. CEREBRAL HEMISPHERES
2. SULCUS
3. CEREBELLUM
4. MEDULLA OBLONGATA
5. GYRI

FIGURE 24.1
DORSAL VIEW OF SHEEP'S BRAIN

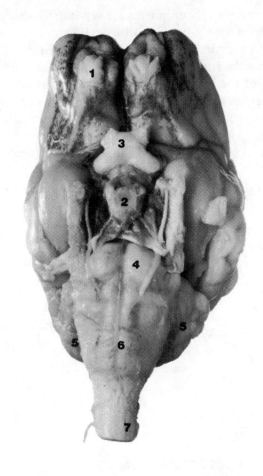

1. OLFACTORY BULB
2. PITUITARY GLAND
3. OPTIC CHIASMA
4. PONS
5. CEREBELLUM
6. MEDULLA OBLONGATA
7. SPINAL CORD

FIGURE 24.2
VENTRAL VIEW OF SHEEP'S BRAIN

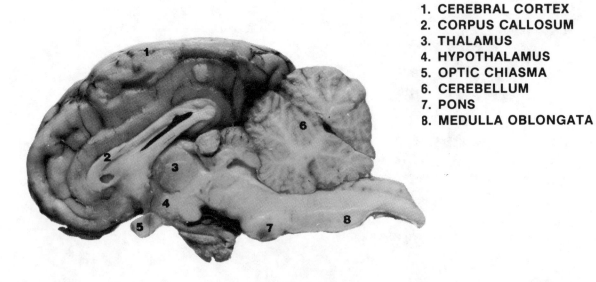

1. CEREBRAL CORTEX
2. CORPUS CALLOSUM
3. THALAMUS
4. HYPOTHALAMUS
5. OPTIC CHIASMA
6. CEREBELLUM
7. PONS
8. MEDULLA OBLONGATA

FIGURE 24.3
SAGITTAL VIEW OF SHEEP'S BRAIN

VIII. **Statoacoustic Nerve (or Auditory):** Origin from the side of the medulla and passes to the inner ear to (1) Organ of Corti in cochlea; (2) semicircular canals. Function: sensory — (1) hearing, (2) equilibrium.

IX. **Glossopharyngeal Nerve:** Origin from the side of the medulla and goes to the muscles and membranes of the pharynx, and to the posterior 1/3 of the tongue. Function: mixed —motor; controls motion in pharynx and tongue — sensory; receives sensations of taste and touch.

X. **Vagus Nerve:** Arises from the side and floor of the medulla and passes to the lungs, heart, stomach, intestine, esophagus, pharynx, and vocal cords. Function: mixed — motor; pharynx, vocal cords, lungs, esophagus, stomach, heart (inhibits heartbeat) — sensory; vocal cords and lungs.

XI. **Spinal Accessory Nerve:** Origin from the side of the medulla and innervates the muscles of the shoulder, palate, larynx, vocal cords, and neck. Function: motor; muscles of pharynx, larynx, and neck.

XII. **Hypoglossal Nerve:** Origin from the ventral region of the medulla and innervating the muscles of the tongue and neck. Function: motor; movement of tongue and neck.

VI. Spinal Cord and Spinal Nerves

Look at a slide of a cross section through the spinal cord under low power. The interior contains a butterfly-shaped area composed of **gray matter** (masses of neuron dendrites and cell bodies). The gray matter is surrounded by **white matter** (masses of axons carrying impulses up and down the cord). The white matter in living material appears lighter in color than the gray matter because of the insulating **myelin sheaths** which surround the axons but not the dendrites and cell bodies. Observe the gray matter under high power and locate some of the neuron cell bodies. The small hole in the center of the cord is the **central canal** which is continuous with the ventricles of the brain.

Connected to the spinal cord are pairs of spinal nerves (33 pair in the pig, 31 pair in humans). Use Figure 24.5 and look again at the dissected fetal pig to locate the spinal cord and the spinal nerves coming from it. There are 8 Cervical, 14 Thoracic, 7 Lumbar, and 4 Sacral pairs of spinal nerves. Note the tiny spherical **dorsal root ganglia** located on the dorsal root of the nerves just before they enter the spinal cord (these will not be visible on each nerve since they are easily destroyed during dissection). There are two enlargements on the spinal cord in the region of the appendages — the anterior one is the **brachial** enlargement and the posterior one is the **lumbar** enlargement. Many of the spinal nerves join to form **plexuses.** The two most prominent ones, shown in Figure 24.5 are the **brachial plexus** (from which nerves extend into the forelimbs) and the **lumbosacral plexus** (from which nerves extend into the hind legs).

VII. Human Reflexes

A.   Reflex Arc

Some stimuli cause rapid responses that come about involuntarily, and often without the intervention of the higher centers of the brain, although in many cases, those centers are informed and may modify the response. These automatic responses are called **reflexes.** The

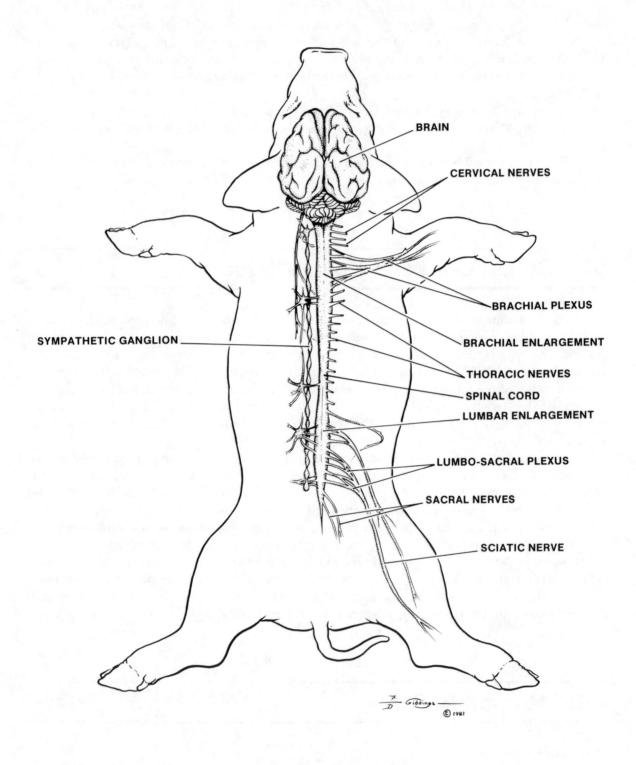

BRAIN

CERVICAL NERVES

BRACHIAL PLEXUS

BRACHIAL ENLARGEMENT

THORACIC NERVES

SPINAL CORD

LUMBAR ENLARGEMENT

LUMBO-SACRAL PLEXUS

SACRAL NERVES

SCIATIC NERVE

SYMPATHETIC GANGLION

FIGURE 24.5 — FETAL PIG NERVOUS SYSTEM, DORSAL VIEW

sensory impulses that initiate a spinal reflex enter the spinal cord through the dorsal root; within the cord, they usually synapse with accessory neurons, which in turn transmit the impulses to motor neurons, and then to some effector, such as a muscle. The whole pathway, from sense organ through the CNS to effector, is called a **reflex arc.** Reflex activity may involve different levels of the central nervous system, depending on the complexity of the reflex. The complexity increases as more neurons become involved in the reflex.

To the right is a diagram of a reflex arc with association neurons. The sensory neuron synapses with several association neurons in the gray matter of the spinal cord. Some of these association neurons may synapse directly with motor neurons on the same side, but some cross to the other side of the cord and there synapse with the other motor neurons and with the additional association neurons that run in **ascending tracts** through the white matter of the cord to the brain.

Examination of specific reflexes such as the patellar reflex is used diagnostically as an indication of normal nervous system function.

For each of the reflexes listed, indicate which structure is the **receptor,** (i.e., receiving the stimulus) and which structure is the **effector** (i.e., causing the response). Be aware of the neural pathway between the two.

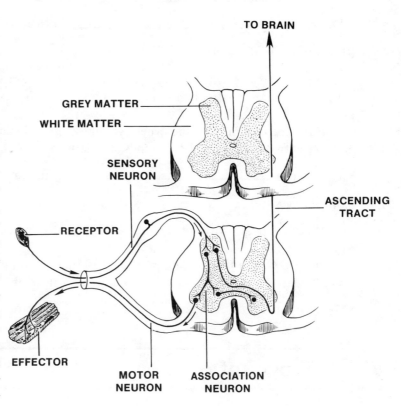

FIGURE 24.6 — THE REFLEX ARC

B.   Patellar Reflex (Knee Jerk)

Have the subject sit on a table so that his legs, from the knee down, hang freely from the table edge. Use a rubber mallet to tap the patellar ligament just below the knee. Note any movement of the lower leg. Test the opposite leg.

Repeat under each of the following conditions, noting how the reflex is affected in each case.

Subject willfully trying to stop the reflex:

Subject tightly clenching fists:

Subject adds a column of figures:

Receptor: _____   Effector: _____

257

C. Achilles Reflex (Ankle Jerk)

Have the subject sit on a chair, his feet hanging free over the edge of the chair. The tester should then place palm of hand under foot and hold it parallel to the floor. Then with the other hand, tap the Achilles tendon and note movement of the foot as it presses into hand of tester.

Have the subject pull against the back of the chair and repeat the test. Result?

Receptor: _____ Effector: _____

D. Photo-Pupil Reflex

Have the subject sit facing a bright light (goose-neck lamp), with eyes closed and covered with hands. After a minute, have him uncover and open one eye. Note the size of the pupil.

Have the person open the other eye, but keep it shielded from the direct light with his hand in between. Observe both pupils. Explain the difference in size.

Receptor: _____ Effector: _____

E. Convergence Reflex

Observe the position of the eyeballs while the subject is looking at an object 20 feet away.

Now have him focus on a pencil held 10 inches from his face. Does the position of the eyeball change as the gaze shifts from a far to a near object? How?

Receptor: _____ Effector: _____

VIII. Frog Reflexes

As you examine some reflexes in a frog do not forget that spinal reflexes, though they do not require the intervention of the brain, may nevertheless be modified by it, and that most reflexes are not nearly as simple as we have outlined above.

A. Reflex Involving Relatively Simple Response

The instructor will prepare a frog in which the cerebral hemispheres have been destroyed by cutting off the cranium (upper part of the head) just posterior to the eyes. Such an animal is called a "decerebrated" frog and is incapable of sensations such as pain. Any response of muscles to external stimuli in such a frog are reflexes. Suspend the frog by the jaw from a clamp on a ring stand, as pictured, so the legs hang freely. Bring a small beaker containing 1% hydrochloric acid up under one foot so that

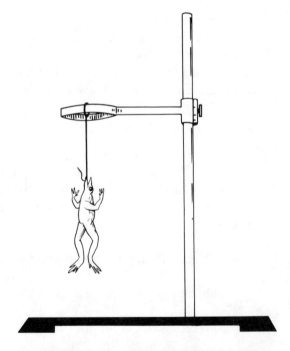

FIGURE 24.7 — FROG PREPARATION

one toe of the frog contacts the acid. Be sure that no other part of the foot or leg touches the sides of the beaker.

Describe what happens.

Submerge the foot in a large beaker of 2% sodium bicarbonate solution for a few seconds, to neutralize the acid, then rinse the foot in a large beaker of tap water.

B. Reflex Involving Complex Purposeful Response

After the frog has rested for several minutes and the legs have relaxed, place, on the upper part of the thigh, a piece of filter paper (about 1/4 in. square) which has been moistened with 30% acetic acid.

Describe the frog's reaction.

Dip the entire leg of the frog in the 2% sodium bicarbonate solution and rinse with tap water. After the legs are relaxed, repeat the experiment, but this time hold the toes of the leg on which the paper is placed, so that the frog cannot move that leg.

Describe what happens.

How did the nerve impulses reach the muscles in the leg which responded, but which was not stimulated by the acid?

C. Effect of Strength of Stimulus on Reflex Time

After the frog has rested for several minutes and the legs are relaxed, suspend the preparation as before and perform the following experiment.

Submerge the long toe of the frog in 0.05% HCl solution. Carefully measure with a second hand of a watch or clock, the time between the application of the acid and the reflex response, if any. (It may take as long as 1½ min.)

Time: _____

After rinsing the toe in a beaker of 2% sodium bicarbonate, dry the toe and repeat the above procedure using 0.25% HCl.

Time: _____

Repeat using 0.5% HCl.

Time: _____

Repeat using 1% HCl.

Time: _____

Plot your data on the graph at the right.

Explain your results.

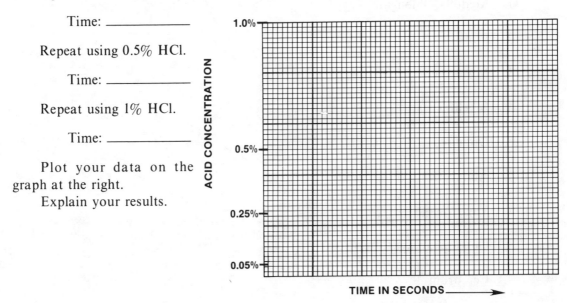

Does an increase in the strength of the stimulus cause the nerve impulses to travel with greater or lesser speed over the nerve fibers?

Why was reflex time different with the higher concentrations of acid?

IX.   Review Questions

A. What region of the brain has, as one of its functions, the coordination of muscular activity?

_____

B. Define these terms:

1. receptor —

2. effector —

3. reflex —

4. stimulus —

5. ascending tract —

6. gyri —

7. central nervous system —

8. peripheral nervous system —

C. Give the **functions** of the following:

1. Medulla Oblongata —

2. Pons —

3. Motor neuron —

4. Myelin sheath —

5. Optic nerve —

6. Abducens nerve —

7. Vagus nerve —

8. Glossopharyngeal nerve —

9. Cerebrum —

10. Thalamus —

D. Describe the sequence of events in a simple reflex arc.

SENSORY MECHANISMS

exercise 25

I. Objectives

After completion of this exercise, the student should be able to do each of the following:

A.   Describe the effects of stimuli on those sense organs capable of responding to them.

B.   Label the major structures found in the various layers of the skin and know the function of each.

C.   Identify taste buds under the microscope.

D.   Describe the functions of the major parts of the human ear and eye.

E.   Label the parts of the human ear and eye.

F.   Be able to diagram, label and explain the functioning of the retina.

G.   Define and explain the causes of **afterimage, nearsightedness, farsightedness, colorblindness,** and **astigmatism.**

H.   Answer the review questions at the end of this exercise.

II. Introduction to Sensory Mechanisms

The nervous system functions in coordinating body activities in response to changes in internal and external stimuli. The particular part of the nervous system specialized in detecting these environmental changes **(stimuli)** is the **sensory receptor,** which initiates the transmission of neural impulses. Receptors vary from simply being "free" nerve endings (pain receptors) to being organized in complex sense organs (eye, ear). The actual sensation detected by the individual is not so much a result of the particular type of stimulus as it is a result of the specific region of the brain to which the impulses are sent.

Although it is said that man has five senses **(sight, taste, hearing, smell,** and **touch),** many more than five types of stimuli can be interpreted by the brain. Stimuli arising from the external environment are detected by the group of receptors known as **exteroceptors.** This group includes the eyes, the auditory part of the ear, the olfactory epithelium, and the receptors located in the skin. Stimuli arising within the body are detected by the group of receptors known as **interoceptors.** This group includes numerous receptors located in the visceral organs (giving rise to sensations of pain, hunger, thirst, nausea, etc.) and in the muscles, tendons, and joints (giving rise to sensations concerned with body position and movement).

Certain receptors display a phenomenon known as **adaptation.** Under continuous stimulation, these receptors respond less and less to a particular stimulus, thus enabling the individual to "get used to" a specific odor, temperature, pressure, etc.

In this exercise, you will learn the basic structure of certain sensory receptors. In addition, a number of experiments will be done on the sensory structures to help you understand their functioning more clearly.

III. Cutaneous Senses

A. Skin Structure

The skin is an organ composed of three basic layers: epidermis, dermis, and subcutaneous layer. The **epidermis** is a region in which mitosis occurs continuously to produce more layers of squamous epithelium, as the outermost layer is being sloughed off continuously. The **dermis** is a region composed mainly of connective tissue. It is in this area that such structures as sweat glands, lymph and blood vessels, hairs (with associated arrector muscle and sebaceous glands), sensory receptors, and nerves are located. The **subcutaneous layer** is the region composed of loose connective tissue that binds the rest of the skin to the body. Here are located numerous fat cells, blood vessels, lymph vessels, the bases of hair follicles, nerves, and deep pressure receptors. Using your text and the model keys as a guide, find the parts of the skin indicated in the diagram on the model and in prepared slides.

There are structurally distinct cutaneous exteroceptors for **pain, touch, pressure, warmth,** and **cold** located in the skin. Stimulation of any one of these will generally give rise to a specific sensation; simultaneous stimulation of different types of receptors will give rise to sensations such as itching, crawling, burning etc.

B. Skin Sensations

Working with your lab partner, perform the following experiments and answer the associated questions.

1. Two-Point Discrimination Test

The distribution of **tactile** (touch) receptors varies over different areas of the body. With two clean toothpick points, touch the skin on the following suggested areas of the body. Start with the points very close together and increase the distance between the points until the person is able to discern two distinct points. Be sure to lift the points off the skin when increasing the distance between points. Record this distance in millimeters for each of the areas tested.

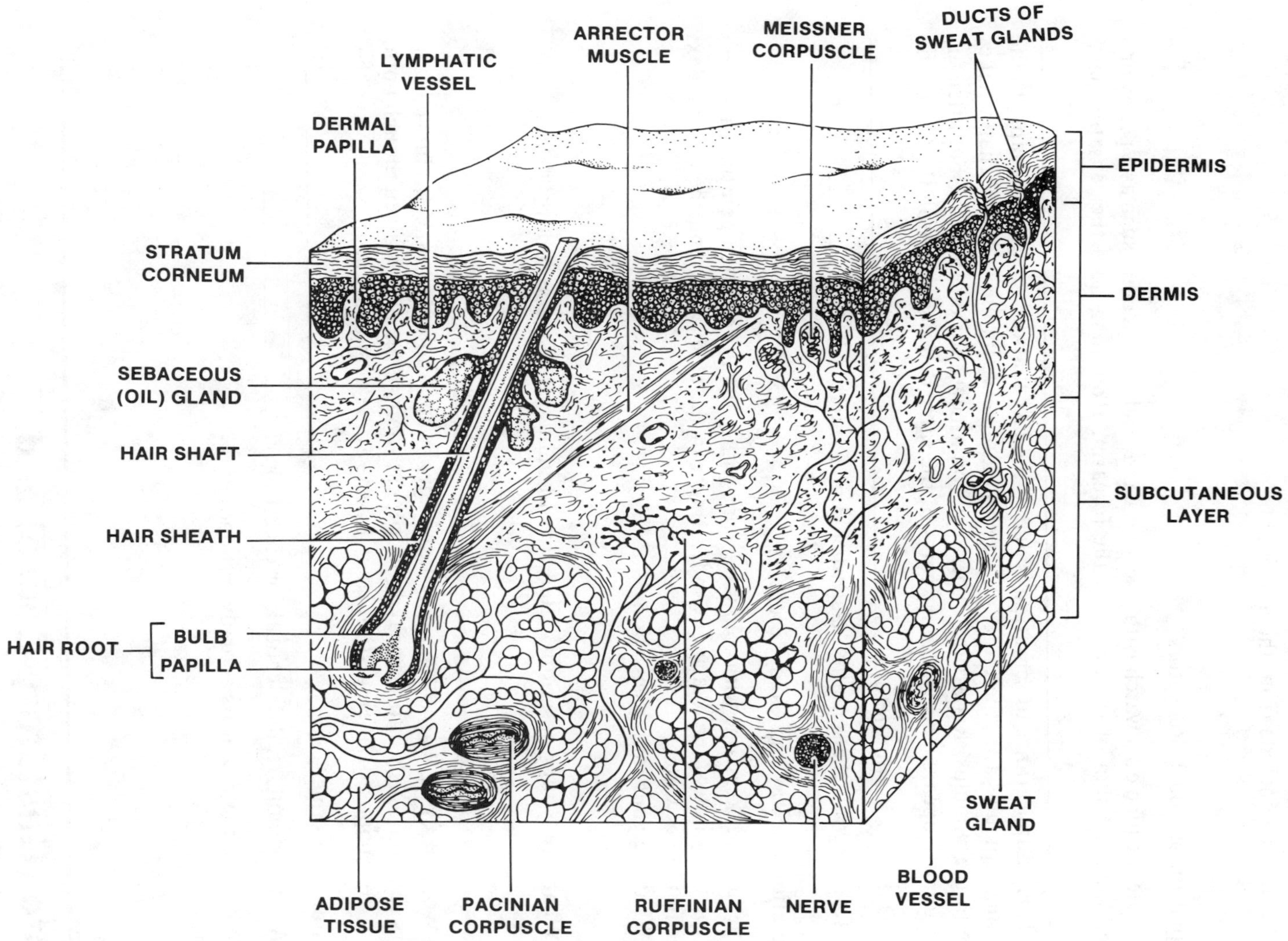

FIGURE 25.1
SECTION OF HUMAN SKIN

EPIDERMIS

DERMIS

SUBCUTANEOUS LAYER

DUCTS OF SWEAT GLANDS

MEISSNER CORPUSCLE

ARRECTOR MUSCLE

LYMPHATIC VESSEL

DERMAL PAPILLA

STRATUM CORNEUM

SEBACEOUS (OIL) GLAND

HAIR SHAFT

HAIR SHEATH

HAIR ROOT

BULB

PAPILLA

SWEAT GLAND

BLOOD VESSEL

NERVE

RUFFINIAN CORPUSCLE

PACINIAN CORPUSCLE

ADIPOSE TISSUE

263

Tip of Nose_____ Back of Neck_____ Palm of Hand_____

Tongue_____ Back of Hand_____ Ball of Thumb_____

Lips_____

What can you conclude from the above data?

2. Adaptation of Touch Receptors

Note the time on a watch with a second hand, and place a coin, such as a penny or a dime, upon the skin of the inside of the forearm. How long does the sensation last?

_____

This disappearance of sensation is due to the fact that the receptors "adapt" to the particular stimulus and, therefore, initiate no more nerve impulses until a change takes place in the stimulus. Repeat the experiment on a new spot on the skin, but after the sensation disappears, add two more coins of the same denomination on the top of the

coin. Did the sensation return? _____

How long did the sensation last when 3 coins were used? _____

3. Adaptation of Heat and Cold Receptors

Set up 3 beakers, one with cold water, one with water at room temperature, and one with tolerably hot water. Place one index finger in the cold water and the other index finger in the hot water.
What are your initial sensations?

What are your sensations after several minutes (fingers still in the beakers)?

IMMEDIATELY after removing fingers from hot and cold beakers, plunge both fingers **simultaneously** into the beaker of water at room temperature. Even though both fingers are supposedly being stimulated by the same temperature of water, are the

sensations the same? _____

Are the stimuli really identical for each finger? _____

What are the receptors actually responding to?

IV. Taste (Gustatory Sense) and Smell (Olfactory Sense)

A. Taste Bud Structure

The **taste buds** are barrel-shaped structures that open through the surface of epithelial cells by way of pores. Within the taste bud are the actual receptor cells with hairlike extensions that detect the materials that are dissolved in saliva. The taste buds are most numerous on the tongue, specifically on some of the **papillae** ("bumps") of the tongue surface. The papillae are rounded mounds, each surrounded by a deep groove. It is into these grooves that the pores of the taste buds open.

Examine a slide of a taste bud. Make a diagram of what you see and label all the parts you can identify. Use the space provided for your diagram.

DIAGRAM OF HUMAN TASTE BUD

B. Olfactory Receptors

The **olfactory receptors** are specialized nasal epithelial cells covering about a postage-stamp-sized area in the upper region of each nasal cavity. The receptors themselves are cells containing hair-like extensions stimulated by molecules dissolved in the mucous covering the nasal epithelium.

C. Distinguishing Between Sensations

Sensations arising from stimulations of taste and smell receptors are often difficult to distinguish from each other. Some of the following experiments are designed to illustrate the relationship between gustatory and olfactory sensations.

1. Stimulation of Taste Buds

Dry the upper surface of your tongue with a clean tissue and have your lab partner place a few granules of sugar on it. Can you taste the sugar immediately? Why?

2. Combined Effects of Taste and Smell

Obtain several pieces of distinctly flavored chewing gum. Dry the surface of your tongue, hold your nostrils shut, and **close your eyes.** Have your partner place a piece of gum (flavor unknown to you) on your tongue. Can you identify the flavor? _____ Now, still holding your nostrils shut, chew the gum. Record your sensations. Open your nostrils and record the changes in sensation.

Nostrils closed:

Nostrils opened:

Dry the surface of your tongue and **shut your eyes,** but do not pinch your nostrils shut. Have your lab partner place one flavor of gum on your tongue while holding a piece of different flavor under your open nostrils. Record your sensations.

3. Combined Effects of Taste, Smell, and Touch

Obtain small ¼ inch cubes of carrot, onion, potato, and apple. Dry your tongue, shut your eyes, and pinch the nostrils shut. Have your lab partner place one of the materials on your tongue. Attempt to identify it (1) immediately, (2) after chewing (nostrils closed), and (3) after opening the nostrils. Record your observations on the chart below.

Substance	Observations		
	Immediately	After Chewing	Open Nostrils
Carrot			
Onion			
Potato			
Apple			

V. Auditory and Equilibrium Senses

A. Ear Structure

The ear is a complex sense organ that can be divided into three basic regions: the **external ear,** the **middle ear,** and the **inner ear.** The external ear is composed of the **pinna (auricle)** and the **external auditory canal** which function in funneling sound waves into the ear. The middle ear is composed of the **tympanic membrane** (eardrum) and the **middle ear bones (malleus, incus, stapes).** These four structures function in transmitting sound vibrations across the **middle ear cavity** to the inner ear. The middle ear cavity and the pharynx are connected by the **Eustachian tube.** The inner ear is actually a double organ. The exteroceptor portion **(cochlea)** is stimulated by vibrations coming from the stapes. The stapes vibrates against the **oval window** membrane of the cochlea. The oval window membrane is at the base of the upper canal. As the oval window vibrates, it in turn causes wave formation in the fluid of the upper canal. The membrane of the **round window** moves outward as the membrane of the oval window moves inward, allowing the fluid to oscillate back and forth. This produces movement of fibers in the **basilar membrane** of the central canal, which in turn causes the compression of individual sensory hairs in the **organ of Corti.** The hair cells connect with the **cochlear branch** of the **statoacoustic nerve** which carries the impulses to the brain. It is in the temporal lobe of the cerebrum where the sound discriminations are made.

In addition to hearing, parts of the inner ear are concerned with the sense of balance and equilibrium. Fluid in the **semicircular canals** moves as the position of the head changes, stimulating tufts of sensory hairs at the bases of the canals. The overall position of the head when still is detected by hair cells in two chambers of the **vestibule,** at the junction of the semicircular canals and the cochlea. Small crystals of calcium carbonate called **otoliths** entangled in these hairs exert unequal pull on the hairs in response to gravitational pull. Nerve impulses from the semicircular canals and vestibule travel to the brain through the

vestibular branch of the statoacoustic nerve. Locate each of the parts labelled in Figure 25.2 on the model of the human ear.

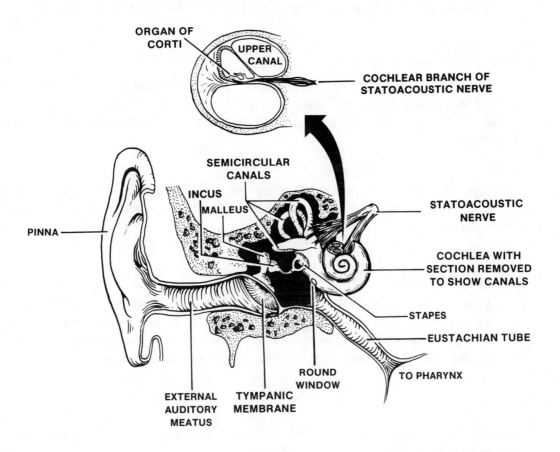

**FIGURE 25.2
THE HUMAN EAR**

B. Experiments on the Ear

1. Visual Sensations in Equilibrium

    Normal maintenance of equilibrium is dependent on sensations from a number of receptors, including those of the semicircular canals, those of muscles and joints, and those associated with vision. The role of vision can be demonstrated by performing the following experiment. With eyes open, stand up straight and raise one foot about 12 inches off the floor. To check the stability of your stance, have a lab partner determine approximately how many inches you sway from side to side. Close your eyes and repeat the procedure. Again have a lab partner check the amount of sway. What differences are noted?

2. Air Pressure in the Middle Ear

    Close your nostrils and mouth and swallow hard. Describe what sensations you experience in your ears.

What is the function of the Eustachian tubes?

Why should there be equal pressure on both sides of the tympanic membrane?

3. Bone Conduction of Sound Waves

Wash and dry the handle of a tuning fork. Cause it to vibrate by tapping the prongs against the palm of the hand. Hold the fork close to the ear until the sound is nearly inaudible, then place the handle firmly between the teeth. Can you hear the sound now? Explain.

Again set the tuning fork to vibrating and place the handle on top of the head. Where does the sound seem to originate? Now close one ear. Where does the sound seem to come from now?

4. Auditory Adaptation or Fatigue

Place the earpieces of a stethoscope in your ears. By pinching the tube, close the passage that leads to your left ear. Sound a tuning fork close to the open end of the stethoscope. When the sound has become almost inaudible to the right ear (through the open tube), open the pinched tube leading to your left ear. What happens? Explain.

VI. Visual Sense

A. Eye Structure

The optical system of the eye consists of a lens system (cornea and a crystalline lens) and a light-sensitive surface (retina). The function of the lens system is to focus light from objects so that the light-sensitive surface is stimulated by the pattern of light and dark areas of the image. The light sensitive surface converts the image pattern into nerve impulses to be transmitted to the brain by way of the optic nerve. Locate each of the parts labelled on Figure 25.3 on the model of the human eye.

B. Dissection of Sheep Eye

Obtain a preserved sheep's eye and find the stumps of the optic nerve and the eye muscles. Then examine the eye and identify the **sclera,** the white cartilaginous coat, covering the hole outside. Note the cloudy area in the front, which is the **cornea** (a specialized area of the sclera that is transparent when living). Identify the **pupillary opening** in the pigmented **iris,** and the **lens,** partially visible through the pupil. Carefully bisect the eye with a sharp scalpel, *in a plane parallel to the front surface,* so that all of the cornea, iris, and lens will be in one half and the back portions of the eye in the other half. The jellylike material filling the main cavity of the eye is the **vitreous humor,** which aids in maintaining eye shape. Examine the back half from the inside. The pale, loose tissue is the **retina.** Note the central **blind spot,** where the nerve fibers from the retina join the **optic nerve** and where the

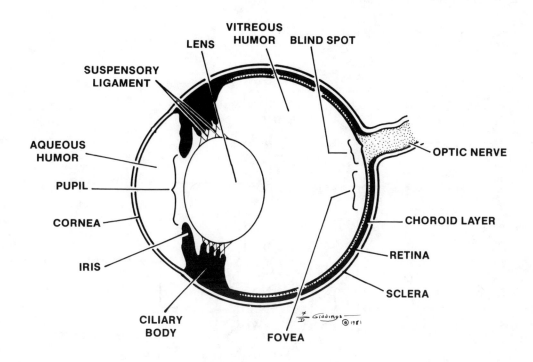

FIGURE 25.3
THE HUMAN EYE

retinal blood vessels pass into the eye. Lift the retina and examine the underlying **choroid coat,** pigmented black. Note the shiny, greenish material covering part of the choroid coat. This is the **tapetum lucidum,** which acts as a mirror reflecting incident light back into the retina. A tapetum is not present in man. Describe the position of the tapetum. Scrape off some of the choroid layer and note that the choroid lies to the inside of the sclera, the outer eye covering already identified above.

Examine the front half of the eye from the inside. Identify the **lens** and note its attachment to the **suspensory ligaments.** The latter are contained in the fine, radially striated membrane which surrounds the lens. The suspensory ligaments are attached peripherally to the **ciliary body,** a ring-shaped muscle pigmented black on its surface. Remove the lens and look through it. Is the curvature of the lens surface the same in front and back? Follow the choroid coat toward the front and note that it continues forward as the **iris.** Probe through the pupil into the anterior chamber containing the **aqueous humor,** a space bounded in front by the cornea.

C.   Structure of the Retina

As was mentioned above, the retina is the layer of the eye which contains the receptor cells responsible for initiation of neural impulses stimulated by light. Looking at Figure 25.4, note that the retina consists of two layers: an outer thin **pigment layer** adjacent to the choroid layer and an inner **neural layer** adjacent to the vitreous humor.

The pigment layer is composed of the tips of the receptor cells, the **rods** and **cones.** The remainder of these cells lie in the neural layer and synapse with **bipolar cells,** which in turn synapse with **ganglion cells.** The axons of the ganglion cells come together at the back of the eyeball to form the optic nerve. Because of this arrangement, with the pigment layer beneath the neural layer, light must pass through the ganglion and bipolar cells to reach the pigment layer and generate an impulse.

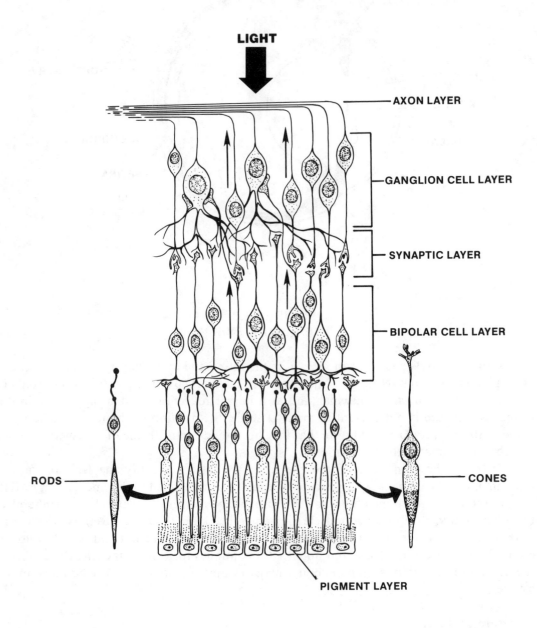

FIGURE 25.4
THE HUMAN RETINA

D. Experiments on the Functioning of the Eye

1. Blind Spot

   With the left eye closed, hold the following black cross and dot about 20 inches from your face with the cross directly in front of the right eye. You should be able to see both figures even though you are looking at the cross. Keep the left eye closed and slowly bring the page closer to the face while continuously looking at the cross with the right eye. At a certain distance the dot will disappear from view, because its image is falling on the blind spot of the eye. Continue to move the page toward the eye. Does the dot reappear? Explain.

   Repeat the experiment using the left eye; this time concentrate on the dot instead of the cross. Are results similar to those for the right eye?

   What is the blind spot of the eye?

2. Visual Acuity Test

   Using the Snellen Chart, determine which line of letters you can easily recognize from the standard distance of 20 feet from the chart. Have your lab partner work with you to note which line you can read accurately. In writing the results of this test, set up the following equation:

$$\frac{20 \text{ (standard distance from the chart)}}{\text{—— (the line you can read from 20 ft. away)}}$$

   For example, if, with your right eye, you can read the line labelled as 40 ft., your vision would be 20/40 for your right eye (i.e., at 20 ft. your right eye can see what the normal-visioned individual can see at 40 ft., indicating poorer than average vision).

   Test each eye separately (without glasses first and then with, if you wear them). Record your results below.

   Right Eye_____ Left Eye _____

   Most cases in which the vision is poorer than average result from structural defects in the eye. If the eyeball is too short, or if the lens is too flat, the incoming light rays will focus at a point behind the eyeball. This is known as **hyperopia,** or **farsightedness,** because the person can see distant objects more clearly than those which are close up. If the eyeball is too long, or if the lens is too convex, **myopia,** or **nearsightedness,** results because the focal point falls in front of the retina. Another common defect, **astigmatism,** occurs as a result of unequal curvature of the cornea or lens. In this condition light rays focus in a line rather than in a point.

3. Colorblindness and Astigmatism

   Examine the charts available to test for astigmatism and colorblindness. Do you have either of these conditions?

What type of colorblindness do you have?

4. Demonstration of Negative Afterimage

Color with bright green, bright yellow, and black crayons or felt pens the American flag pictured below. Place the colored flag at a distance of approximately one foot in front of you in a strong light and concentrate your gaze at the small target "X" in the center of the flag. Keep your eyes focused on the target for at least ten full seconds. Now look at a blank piece of white paper or at the white ceiling of the room. Describe, in the space below, the afterimage you see.

	Green	
	Black	
	Green	X
Black		
Green		
Black		

Afterimages appear to form as a result of fatigue in the receptor cells, and possibly in the bipolar and ganglion cells as well. It is thought that the cells increase their rate of firing in response to one color, but decrease their rate of firing in response to its complementary color. When one stares intently at a uniform green field for a period of time, the cells which fire at an increased rate for green (green on-red off cells) become fatigued, so that if one looks at a uniform white field, the reduced firing of these cells is interpreted as redness.

VIII. Review Questions

A. What are the five cutaneous senses?

1. _____

2. _____

3. _____

4. _____

5. _____

B. In the two point discrimination test, which area had the highest number of exteroceptors?

_____

C. Diagram and label the structure of a taste bud in the space provided below.

D. From your tests with olfactory and gustatory receptors, what are your conclusions regarding the relationship of these two senses?

_____

_____

_____

E. What is auditory adaptation or fatigue?

_____

_____

_____

F. What structure helps to regulate the air pressure on either side of the tympanic membrane?

_____

G. Trace the steps that would be involved in the passage of a beam of light through the eye until the impulse leaves through the optic nerve.

H. Describe the functions of the following:

1. vestibular branch of the statoacoustic nerve

2. round window

3. oval window

4. cochlear branch of the statoacoustic nerve

5. organ of Corti

6. semicircular canals

7. retina

8. crystalline lens

I. What kind of lens would be prescribed to correct farsightedness? _____
   Nearsightedness? _____ Astigmatism? _____

SKELETAL SYSTEM

exercise 26

I. Objectives

After the completion of this exercise, the student should be able to do each of the following:

A.   Diagram and label the parts of a bone in longitudinal and cross sections.

B.   Identify the major parts of bone tissue under the microscope.

C.   Identify the major bones of the human skeleton (both articulated and disarticulated) as well as in diagrams.

D.   Name and describe the three major types of joints and give examples of each.

E.   Name the various types of freely movable joints and give examples of each.

F.   Identify the various types of teeth in the human.

G.   Diagram and label a longitudinal section through a human tooth.

H.   Describe the functions of the various types of teeth and the layers of the individual tooth structure.

I.   Answer the review questions at the end of this exercise.

II. Introduction to the Skeletal System

The skeletal system in the vertebrate animal is comprised of **bone, cartilage** and **ligaments.** It functions in such things as body support and protection of internal organs, and also provides sites for skeletal muscle attachments, thus functioning in locomotion. In addition, it is involved in the formation of blood cells and serves as a storage site for calcium and phosphorus salts.

In this exercise we will examine the structure of the skeletal system, beginning with bone and cartilage and its microscopic structure and ending with the gross anatomy of the entire system. Various types of joints (sites where two bones meet) and their movements will also be studied.

III. Bone Structure

A. Parts of a Bone

Bones are comprised of a type of connective tissue in which the **osteocytes** (individual cells) are separated by an intercellular **matrix** hardened by the deposition of tricalcium phosphate $(Ca_3(PO_4)_2)$ and calcium carbonate $(CaCO_3)$. This bony tissue exists as two types: a solid **compact** type and a **cancellous** type (porous spongy type). In long bones, the **diaphysis** (shaft portion) is primarily a cylinder of compact bone surrounding a central cavity containing **yellow bone marrow.** The outer surface of the diaphysis is covered by a tough fibrous membrane known as the **periosteum.** The **epiphysis** (enlarged end) of long bones is primarily spongy bone whose spaces contain red bone marrow. The spongy bone is covered by a thin layer of compact bone, which is covered in turn by **articular cartilage.** In young animals there is a thin layer of cartilage (**epiphyseal disc**) between the diaphysis and epiphysis; this disc is the site of bone elongation.

B. Slide of Ground Bone

The structure of spongy bone is such that the osteocytes can easily acquire their nutrients and eliminate their wastes via the blood capillaries in the numerous marrow-filled spaces around the bony network. In compact bone, because of its density, blood vessels from the periosteum are distributed through the bony tissue in passageways known as **Haversian canals.** These canals contain blood and lymph vessels, as well as nerve fibers. The osteocyte in its **lacuna** (cavity) communicates with the Haversian canal and neighboring osteocytes by even smaller canals known as **canaliculi,** into which protoplasmic extensions of the bone cell extend. The osteocytes are distributed in concentric rings around a central Haversian canal; the concentric layers of matrix surrounding the canal are known as **lamellae.** The entire structure made up of the Haversian canal, lamellae, and osteocytes is known as an **Haversian system,** a microscopic structure extending lengthwise in compact bone. This structure is continuously being broken down and rebuilt by the body.

Examine a prepared slide of ground bone. Note that the actual osteocytes are no longer present in the preparation. Locate the structures printed in boldface in the previous paragraphs with the aid of Figure 26.1. Is there more than one Haversian system present?

IV. Cartilage and Ligaments

Cartilage is a type of connective tissue in which the cells (**chondrocytes**) are separated by a matrix of collagen or elastic fibers imbedded in a firm gel. This matrix allows cartilage to function as a supportive tissue while still being somewhat flexible. Similar to osteocytes, the chondrocytes are found in spaces called **lacunae,** but they are not arranged in rings connected by canaliculi as in bone.

Cartilage is found at the ends of bones in articulations, the nose, the pinna of the ear, the larynx, the trachea, the intervertebral discs (pads between vertebrae), and in the connections between the ribs and the sternum.

A. Observe a prepared slide of cartilage and identify the parts in boldface type from the description above.

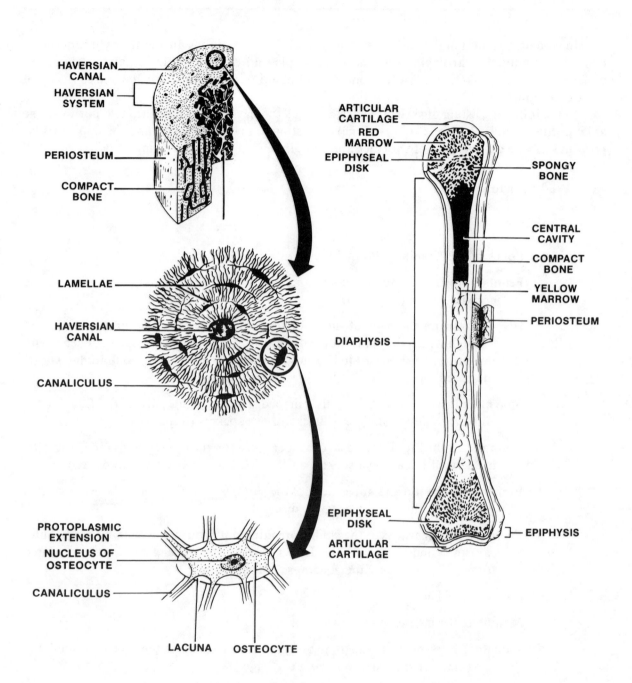

FIGURE 26.1 — BONE STRUCTURE

Ligaments are composed of fibrous connective tissue and function to connect bones together. Fibrous connective tissue is composed of cells sandwiched between layers of tightly packed **collagen** and **elastic fibers.** Thus ligaments have a long, stringy texture. **Tendons** are also composed of fibrous connective tissue, but connect muscle to bone instead of bone to bone.

B.   Locate the cells and the fibers in a prepared slide of fibrous connective tissue.

V. Skeletal Structure

In vertebrates, most of the skeletal system is initially cartilaginous during embryonic development. Eventually, this cartilaginous framework is replaced by bone. At birth, man has about 270 bones making up his skeleton; in adulthood this number is reduced to about 206 due to fusion of individual bones.

In the following listing of skeletal bones, the skeleton has been divided into two portions: the **axial portion** (skull, backbone, and rib cage) consisting of about 80 bones, and the **appendicular portion** (arms, legs, and part of shoulder and hip girdles, all bones occurring in pairs).

A. Axial Skeleton

    1. Skull

        a. Bones of the Cranium (Brain Case)

            Frontal: forms the forehead region; separated from the parietal bones by the **coronal suture** (unpaired)

            Occipital: forms the back of the skull where it rests on the vertebral column, and through which the spinal cord of the nervous system passes as it connects with the brain; it is separated from parietal bones by the **lambdoidal suture** (unpaired)

            Sphenoid: an irregular "butterfly" shaped bone forming part of the floor of the cranium, sides of skull, and backs of eye sockets (unpaired)

            Ethmoid: at the top of the nasal chamber, between the eye sockets, and forms part of the wall of each eye socket, on the side toward the nose (unpaired)

            Parietals: at the top and upper sides of the skull, behind the frontal bone; separated from each other by the **saggital suture**

            Temporals: at the sides of the skull, spreading out around the opening of the outer ear canal, and containing the deeper parts of the hearing apparatus; separated from the parietals by the **squamosal suture**

        b. Bones of the Face

            Mandible: forms the lower jaw (unpaired)

            Vomer: forms part of the thin, bony partition between the nasal cavities, the remainder of the partition being cartilage

            Maxillae: together forming the upper jaw

            Zygomatics: form the upper, bony prominences of the cheeks and the sides and lower borders of the eye sockets; each joins a zygomatic process of the temporal bone to form a zygomatic arch under and over which pass the strong chewing muscles

            Lacrimals: form a small part of the medial walls of the eye sockets

            Nasals: form the hard upper part or bridge of the nose (most of the nasal support is cartilage)

Inferior nasal conchae: curved, shell-like bones sloping inwardly, like shelves, form the lateral walls of the nasal chambers, below the curved **superior** and **middle conchae** which are parts of the ethmoid bone. The conchae are very brittle, thin, delicate bones and, therefore, are often damaged in preparation of skulls; they may be hard to locate if partly destroyed

Palatines: constitute the back part of the hard palate of the roof of the mouth.

Ossicles or ear bones: paired bones in the middle ear cavities (6 bones)

> **Malleus** or "hammer": attached to the inside of the tympanic membrane or eardrum
>
> **Incus** or "anvil": in between the malleus and stapes
>
> **Stapes** or "stirrup": in contact with the oval window membrane

Hyoid bone: a U-shaped bone suspended by ligaments from the pointed styloid processes on the lower part of the temporal bones. (This is not truly a bone of the skull, but it is included here as a matter of convenience.)

On the diagrams of the skull, Figure 26.2 and 26.3, label the bones of the cranium and of the face, using texts and skulls available in lab. (Note: coloring each bone of the diagram a different color will help in distinguishing them.)

2. Vertebral Column (26-33 vertebrae; 24 are moveable; singular: vertebra). See Figure 26.4.

Cervical vertebrae: in the neck region (7 bones). The first cervical vertebra is the **atlas** and supports the head. The second vertebra is the **axis,** which has an **odontoid process** which projects through the ring of the atlas to allow the head to pivot.

Thoracic vertebrae: in the chest region (12 bones)

Lumbar vertebrae: in the region of the "small of the back" (5 bones)

Sacral vertebrae: the "sacrum" which consists of fused vertebrae, usually 5, and forms a triangular wedge of bone in the back part of the pelvic girdle (counted as 1 bone in the adult skeleton)

Coccygeal vertebrae: (Coccyx) 4 or 5 vertebrae, sometimes each separated from the others, but usually fused in various combinations to form the "tail" bone. (May be any number from 1 to 5, inclusive. There is a tendency for more fusion of these bones in the skeleton of the male than in the female.)

3. Thorax (25-27 bones)

Sternum (breastbone): consisting of 3 parts (**manubrium, gladiolus** or body, and **xiphoid**) which become fused in elderly individuals (usually counted as one bone, but may be 3 in younger adults.) The manubrium and gladiolus are named for their resemblance to the handle of a sword.

Ribs or **Costae** (12 pairs)

> **"true ribs":** 7 pairs attached directly to the sternum by means of their cartilages
>
> **"false ribs":** 5 pairs. The first 3 pairs are attached to the cartilages of adjacent ribs rather than to the sternum directly. The last 2 pairs are termed "floating ribs" because they are not attached to the sternum either directly or indirectly.

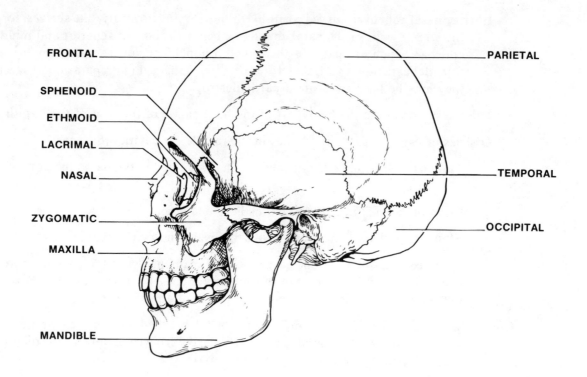

FIGURE 26.2 — HUMAN SKULL (Side View)

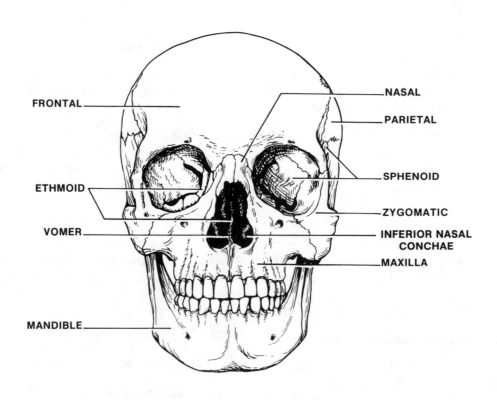

FIGURE 26.3 — HUMAN SKULL (Front View)

B.   Appendicular Skeleton

    1.   Shoulder Girdle

        Scapulae (singular, scapula): shoulder blades

        Clavicles: collar bones

    2.   Upper Appendages (30 bones, each)

        Humerus: upper arm bone

        Radius: lower arm bone, on the thumb side

        Ulna: lower arm bone, on the little finger side; the longer of the two lower arm bones

        Carpal bones: in the wrist (8 bones)

        Metacarpal bones: in the hand (5 bones)

        Phalanges: finger bones; each finger has 3; the thumb has 2 (14 bones)

    3.   Hip Girdle

        Innominate bones: consisting of the fused **ilium** (uppermost and largest portion), **ischium** (strongest portion — directed slightly posteriorly), and **pubis** (superior and anterior to ischium). Counted as 1 bone on each side after the fusion has occurred. The joint between the two pubic bones is called the **pubic symphysis.**

    4.   Lower Appendages (30 bones each)

        Femur: thighbone

        Patella: kneecap

        Tibia: larger, medial lower leg bone

        Fibula: smaller, lateral lower leg bone

        Tarsals: form the ankle (7 bones)

        Metatarsals: foot bones (5 bones)

        Phalanges: toe bones; big toe has 2, other toes have 3 (14 bones)

    On Figure 26.5, the human skeleton anterior and posterior views, label the bones indicated by label lines.

VI. Joints (Articulations)

    Joints are considered to be the junctions between two bones and are classified on the basis of the amount of movement allowed. There are three basic categories:

A.   Immovable joints **(Synarthroses)**

    These are joints in which the bones are connected to each other by fibrous connective tissue. The sutures of the skull and the epiphyseal discs of the long bones are typical examples.

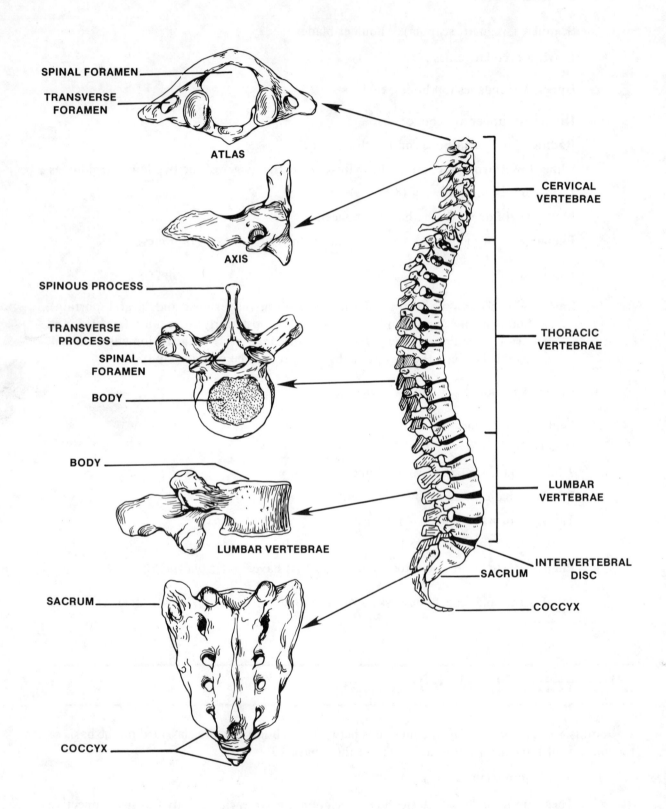

SPINAL FORAMEN

TRANSVERSE FORAMEN

ATLAS

AXIS

SPINOUS PROCESS

TRANSVERSE PROCESS

SPINAL FORAMEN

BODY

BODY

LUMBAR VERTEBRAE

SACRUM

COCCYX

CERVICAL VERTEBRAE

THORACIC VERTEBRAE

LUMBAR VERTEBRAE

INTERVERTEBRAL DISC

SACRUM

COCCYX

FIGURE 26.4 — THE VERTEBRAL COLUMN

THE HUMAN SYSTEMS

INDEX

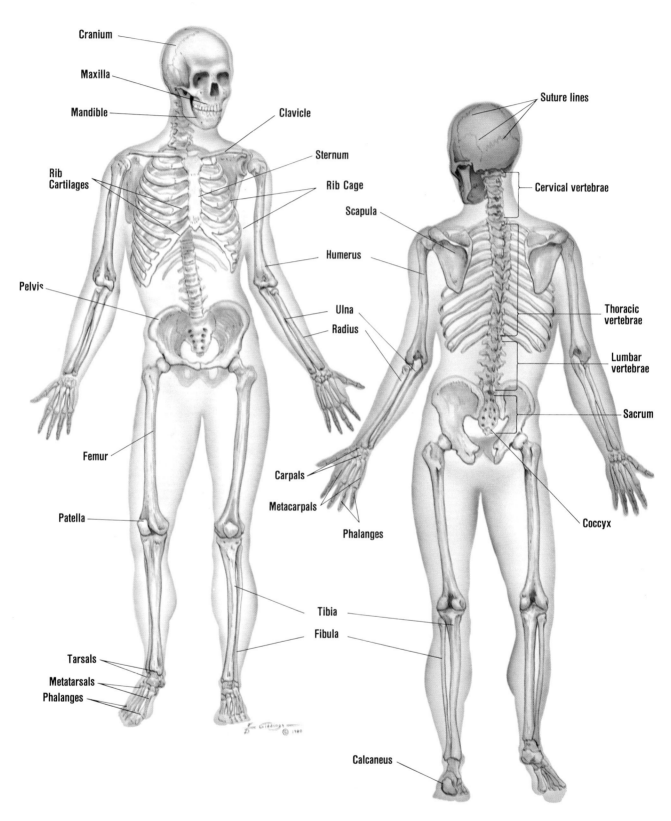

Cranium

Maxilla

Mandible

Clavicle

Sternum

Rib Cartilages

Rib Cage

Scapula

Humerus

Pelvis

Ulna

Radius

Femur

Patella

Carpals

Metacarpals

Phalanges

Tibia

Fibula

Tarsals

Metatarsals

Phalanges

Calcaneus

Suture lines

Cervical vertebrae

Thoracic vertebrae

Lumbar vertebrae

Sacrum

Coccyx

THE SKELETAL SYSTEM

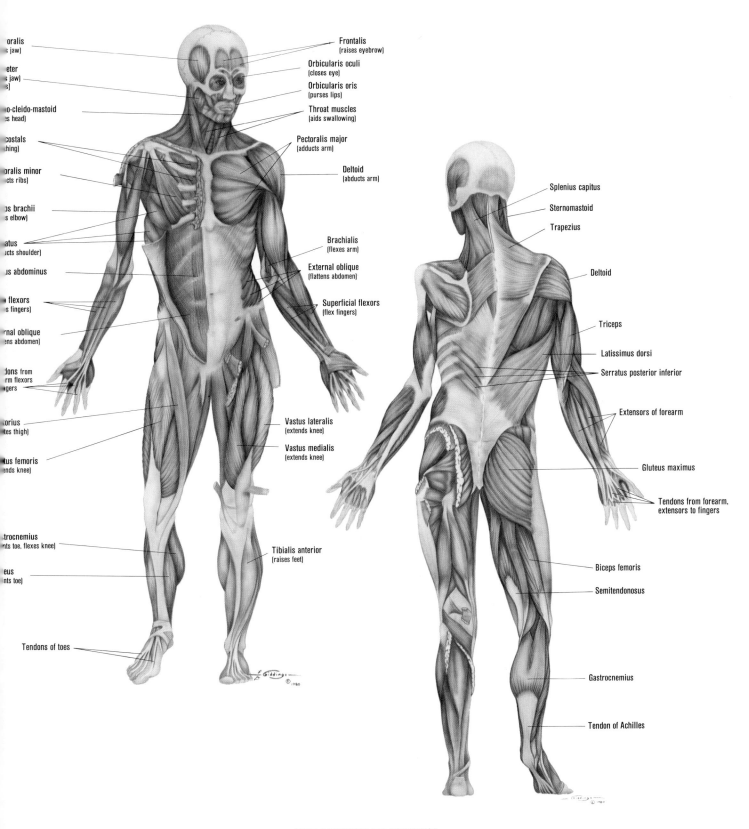

...oralis
(s jaw)

...eter
(s jaw)
...s)

...o-cleido-mastoid
...es head)

...costals
...thing)

...oralis minor
...cts ribs)

...s brachii
...s elbow)

...atus
...cts shoulder)

...us abdominus

...a flexors
...s fingers)

...rnal oblique
...ens abdomen)

...ions from
...m flexors
...gers

...orius
...tes thigh)

...us femoris
...nds knee)

...trocnemius
...nts toe, flexes knee)

...eus
...ints toe)

Tendons of toes

Frontalis
(raises eyebrow)

Orbicularis oculi
(closes eye)

Orbicularis oris
(purses lips)

Throat muscles
(aids swallowing)

Pectoralis major
(adducts arm)

Deltoid
(abducts arm)

Brachialis
(flexes arm)

External oblique
(flattens abdomen)

Superficial flexors
(flex fingers)

Vastus lateralis
(extends knee)

Vastus medialis
(extends knee)

Tibialis anterior
(raises feet)

Splenius capitus

Sternomastoid

Trapezius

Deltoid

Triceps

Latissimus dorsi

Serratus posterior inferior

Extensors of forearm

Gluteus maximus

Tendons from forearm,
extensors to fingers

Biceps femoris

Semitendonosus

Gastrocnemius

Tendon of Achilles

THE MUSCULAR SYSTEM

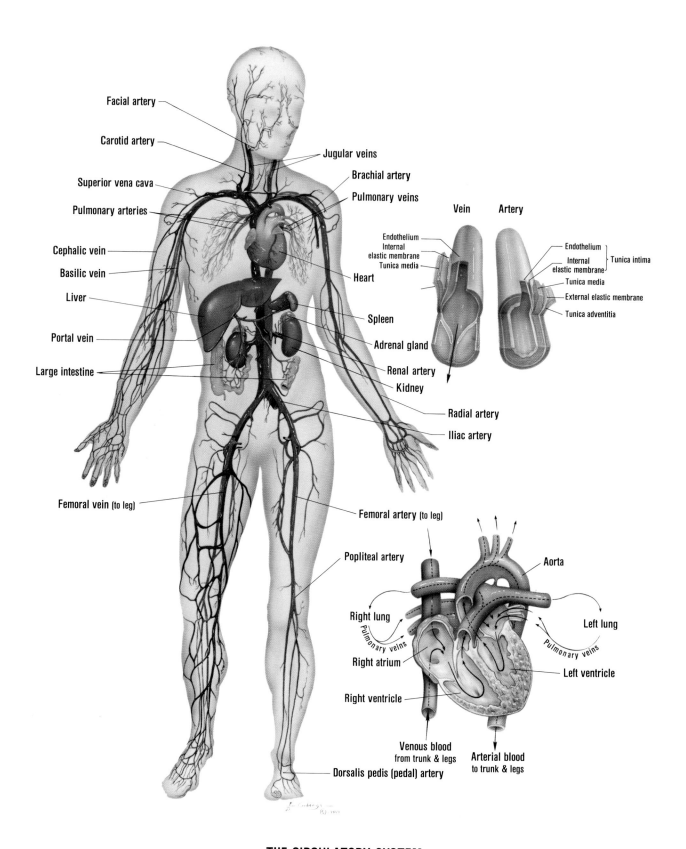

Facial artery
Carotid artery
Superior vena cava
Pulmonary arteries
Cephalic vein
Basilic vein
Liver
Portal vein
Large intestine
Femoral vein (to leg)

Jugular veins
Brachial artery
Pulmonary veins
Heart
Spleen
Adrenal gland
Renal artery
Kidney
Radial artery
Iliac artery

Femoral artery (to leg)
Popliteal artery

Dorsalis pedis (pedal) artery

Vein
Artery

Endothelium
Internal elastic membrane
Tunica media

Endothelium
Internal elastic membrane
Tunica media
External elastic membrane
Tunica adventitia

Tunica intima

Aorta
Right lung
Pulmonary veins
Right atrium
Right ventricle
Left lung
Pulmonary veins
Left ventricle

Venous blood
from trunk & legs
Arterial blood
to trunk & legs

THE CIRCULATORY SYSTEM

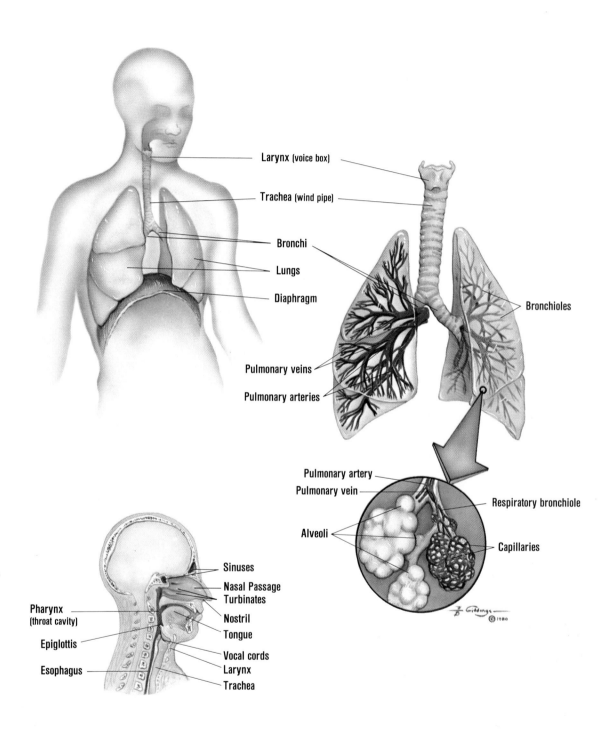

Larynx (voice box)

Trachea (wind pipe)

Bronchi

Lungs

Diaphragm

Bronchioles

Pulmonary veins

Pulmonary arteries

Pulmonary artery

Pulmonary vein

Respiratory bronchiole

Alveoli

Capillaries

Sinuses

Nasal Passage

Turbinates

Pharynx (throat cavity)

Nostril

Tongue

Epiglottis

Vocal cords

Larynx

Esophagus

Trachea

THE RESPIRATORY SYSTEM

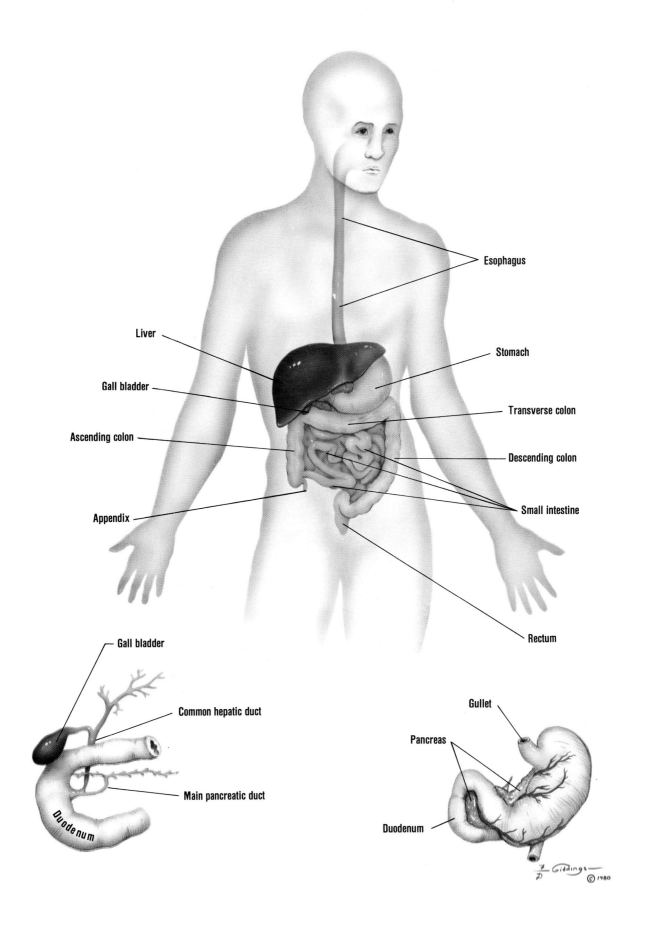

Esophagus

Liver

Gall bladder

Ascending colon

Appendix

Stomach

Transverse colon

Descending colon

Small intestine

Rectum

Gall bladder

Common hepatic duct

Main pancreatic duct

Duodenum

Gullet

Pancreas

Duodenum

THE DIGESTIVE SYSTEM

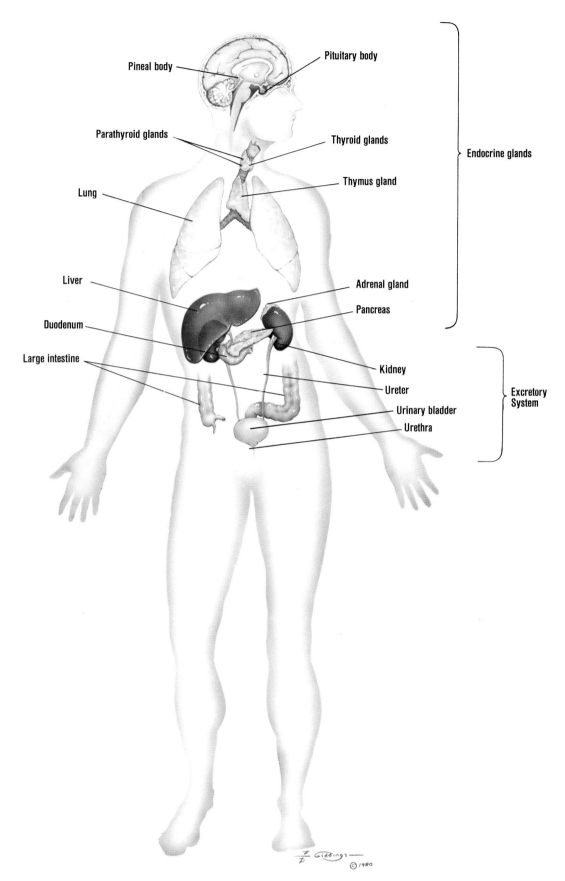

Pineal body

Pituitary body

Parathyroid glands

Thyroid glands

Thymus gland

Lung

Liver

Adrenal gland

Pancreas

Duodenum

Large intestine

Kidney

Ureter

Urinary bladder

Urethra

Endocrine glands

Excretory System

THE URINARY AND ENDOCRINE SYSTEM

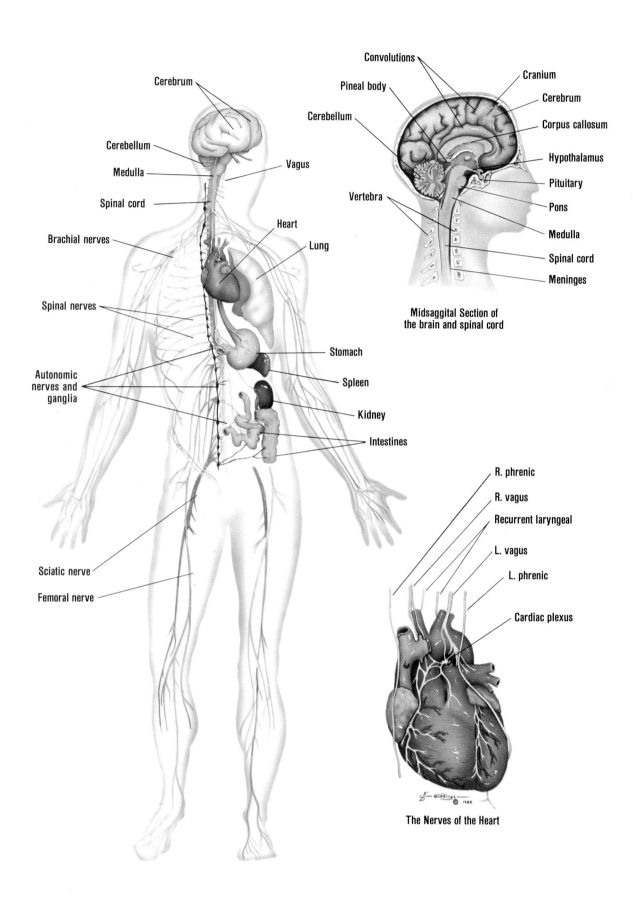

Cerebrum

Cerebellum

Medulla

Spinal cord

Brachial nerves

Spinal nerves

Autonomic nerves and ganglia

Sciatic nerve

Femoral nerve

Vagus

Heart

Lung

Stomach

Spleen

Kidney

Intestines

Convolutions

Pineal body

Cerebellum

Vertebra

Cranium

Cerebrum

Corpus callosum

Hypothalamus

Pituitary

Pons

Medulla

Spinal cord

Meninges

Midsaggital Section of the brain and spinal cord

R. phrenic

R. vagus

Recurrent laryngeal

L. vagus

L. phrenic

Cardiac plexus

The Nerves of the Heart

THE NERVOUS SYSTEM

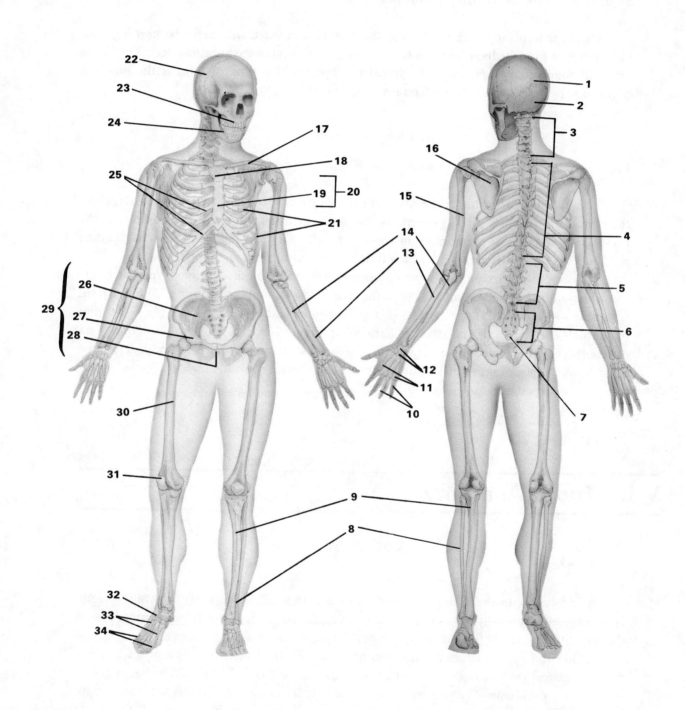

FIGURE 26.5 — HUMAN SKELETON — ANTERIOR AND POSTERIOR VIEWS

Diagram A of Figure 26.6 is a section through a suture. Note that the two bones are separated by **fibrous connective tissue** which is continuous with the **periosteum** on the external surface and with the **dura mater** on the inner surface.

B.   Slightly Movable Joints (**Amphiarthroses**)

These are joints in which the ends of the bones are covered with **articular cartilage** and have a disc of **fibrocartilage** between the two bones which has a cushioning effect. The bones are held together by **ligaments**. A typical example of this type of joint is the junction between vertebrae, as shown in Diagram B of Figure 26.6.

C.   Freely Movable Joints (**Diarthroses**)

These are joints in which the ends of the bones are covered with **articular cartilage** and the bones held together with **ligaments** as above; the distinguishing feature, however, is the presence of a **synovial membrane** on the inner surface of the ligament. This membrane produces a slippery viscous fluid which functions in lubricating the joints.

This is the most common type of joint in the body. Some various types of freely movable joints are: (1) **ball and socket joint** (example — shoulder and hip); (2) **hinge joint** (example — elbow, knee, and ankle); (3) **gliding joint** (example — wrist); (4) **pivot joint** (example — between axis and atlas). Diagram C in Figure 26.6 is of the knee joint, one of six different types of diarthrotic joints. Working with the skeleton or your own body, determine the different types of movements allowed by various freely movable joints.

VII.   Tooth Anatomy

A.   Tooth Structure

Associated with the skeletal system because of their chemical makeup, are the **teeth**. However, embryonically they are derived from ectoderm rather than from mesoderm as is the skeleton. Internally they have the same structure as scales. Examine the lower surface of the maxilla and upper surface of the mandible of a skull and notice the sockets in which the roots of the teeth are located. Usually by the time an individual is 2 years old he has acquired all 20 of his **deciduous** ("baby") **teeth**. At about 6 years of age, the **permanent teeth** begin erupting. The loss of the deciduous teeth arises from a reabsorption of the roots accompanying the growth of the permanent teeth underneath.

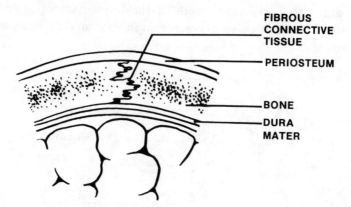

A. SYNARTHROSIS

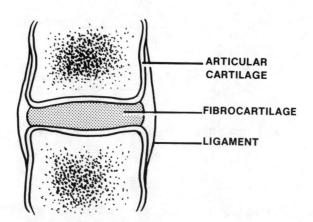

B. AMPHIARTHROSIS

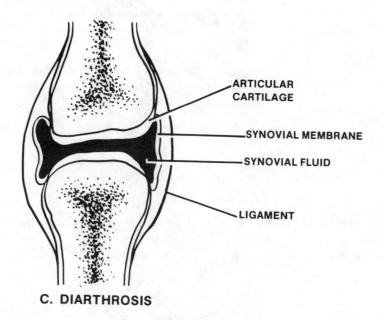

C. DIARTHROSIS

FIGURE 26.6 — TYPES OF JOINTS

Figure 26.7 shows the permanent teeth of the human upper jaw. Locate the parts indicated and be able to identify these structures on the jaw models, head models, and skulls available in the lab.

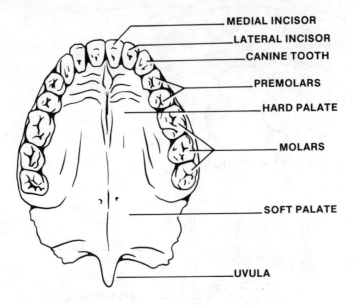

FIGURE 26.7 — HUMAN UPPER JAW

B. Longitudinal Section of a Tooth

Refer to Figure 26.8 to locate the structures given in boldface type on the next page.

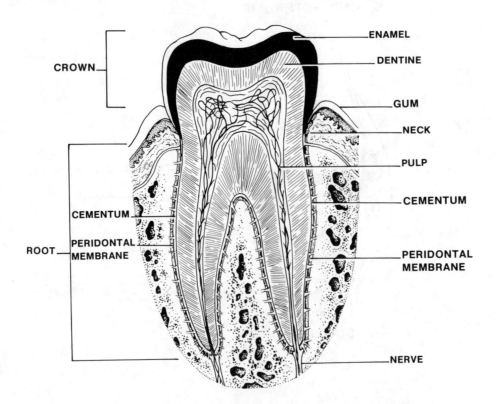

FIGURE 26.8 — SECTION THROUGH A TOOTH

The tooth is comprised of three main regions: the **crown** (above the gum); the **root** (embedded in the **bony socket** of the jaw, which is covered by the fleshy **gum**); and the **neck** (a slightly constricted area between the root and the crown). The hard parts of the tooth include the **enamel, dentine,** and **cementum;** even though these are like bone in the structure, they are much harder. (Why?) **Enamel** is the outer covering of the crown; **cementum,** continuous with the enamel, covers the root; **dentine** makes up most of the tooth (underneath the enamel and cementum).

The soft parts of the tooth are the **pulp** and the **peridontal membrane.** The pulp is the innermost part of the tooth, consisting of connective tissue plus capillaries, lymph vessels, and nerve fibers which enter the pulp through an opening (**foramen**) in the root. The peridontal membrane is a fibrous membrane covering the root and lining the bony socket (here it functions as the periosteum); its function is to anchor the tooth to the socket and to serve as a shock absorber.

Obtain a slide of tooth development. On this slide, the tooth has not yet erupted through the gum.

In the space provided below make a diagram of the tooth on the slide and label the **enamel, dentine, pulp,** and **gum.**

VIII. Review Questions

A. What general functions does the skeletal system perform? _____

_____

_____

B. Is the epiphysis of a long bone the *only* place spongy bone is located? Explain. _____

_____

_____

_____

C.   Identify the parts of the Haversian system as shown on the model in the lab. _____
_____
_____

D.   What does the shape of different teeth indicate about their particular function? _____
_____
_____

E.   Which teeth are considered to be the "wisdom" teeth? _____

F.   How many permanent teeth are there in a complete set? _____

G.   List the 3 major types of joints and give an example of each.
_____
_____
_____

H.   Identify the following terms:

    (1)   Osteocyte —

    (2) $Ca_3(PO_4)_2$ and $CaCO_3$ —

    (3)   Epiphyseal disc —

    (4)   Canaliculi —

    (5)   Ossicles —

    (6)   Axial skeleton —

    (7)   Appendicular skeleton —

    (8)   Ball and socket joint —

    (9)   Atlas and Axis —

I.   What is the function of ligaments? _____
    tendons? _____

J.   List three places in the body in which cartilage can be found. _____
_____

288

MUSCULAR SYSTEM

I. Objectives

After the completion of this exercise, the student should be able to do each of the following:

A. Identify the three different types of muscle tissue under the microscope.

B. Identify the type of muscle in the structures named and the function of the muscle in that structure.

C. Define the terms **flexion, extension, abduction, adduction,** and **rotation** and give an example of each.

D. Recognize on models or diagrams, the fourteen major human muscles mentioned in this exercise.

E. Name the muscles involved in some of the simple activities outlined in this exercise.

F. Define **muscle fatigue, threshold,** and **oxygen debt.**

G. Define the following, and identify each on a physiograph or kymograph tracing:

> **Tetanus** —
>
> **Contraction** —
>
> **Latent Period** —
>
> **Refractory period** —
>
> **Relaxation** —

H. Answer the review questions at the end of this exercise.

II. Introduction

Approximately one half of the body is composed of muscle. This tissue is specialized for contraction, thus allowing for movement of body parts, whether it be in the movement of an arm, leg, emptying of the urinary bladder, churning action of the stomach, etc. Within the vertebrates, muscle tissue has differentiated along three lines; these three resulting types of muscle will be described below. In addition to a study of muscle anatomy or structure, this exercise will include simple demonstrations of muscle physiology or function.

III. Types of Muscle Tissue

Obtain a slide of each of three muscle types. Examine each slide carefully and make a diagram in the spaces provided below.

A. **Skeletal Muscle:** Characteristics:
1. Makes up most of the body muscle mass
2. Attached to bone (usually)
3. Under conscious control
4. Involved in locomotion, movement of bones, and maintaining posture
5. Each muscle fiber has many nuclei (along the edge of the fiber) and many cross striations (lines)

B. **Visceral Muscle, or smooth muscle:** Characteristics:
1. Found in walls of internal organs
2. Not usually under conscious control
3. Involved in contraction of stomach, intestine, blood vessels, lungs, etc.
4. Each muscle cell has one nucleus, located in the center of the fiber; striations are lacking

C. **Cardiac Muscle:** Characteristics:
1. Located only in the heart
2. Not usually under conscious control
3. Branched muscle fibers; one nucleus per cell; cross striations present
4. Intercalated discs (dark lines at junction of 2 cells) present (function in rapid conduction of impulses from one cell to another)

D.   Operation of Specific Organs

Considering the following structures, indicate the type of muscle found in each and the particular role of the muscle tissue in each case.

Structure	Muscle Type	Role
Tongue & Jaw		
Pyloric sphincter		
Diaphragm & muscles between ribs		
Intestinal Wall		
Wall of Urinary Bladder		
Anal Sphincter		
Ventricle of Heart		
Walls of Arteries		
Walls of Uterus		

IV. Muscle Action

Contraction is the *only* action a muscle can perform. When a muscle relaxes, no work is done. Skeletal muscles do not work singly, but rather in groups — with one group being antagonistic to the other (i.e., having an opposite effect). Skeletal muscles can be classified on the basis of the type of movement they produce. There are a large number of different types of movement, but in this exercise we will restrict ourselves to the five listed below.

A.   **Flexion**

This action is a movement of a body part such that the angle of a joint is decreased (e.g., bending the elbow). A flexor muscle is one which bends a limb.

B.   **Extension**

Extension involves a movement of a body part such that the angle of a joint is increased (e.g., straightening of the arm).

C.   **Abduction**

This is movement of a body part away from the median axis of the body (e.g., lifting the arm straight out laterally).

D.   **Adduction**

This constitutes movement of a body part toward the median axis of the body (e.g., lowering the arm to the side of the body).

E.   **Rotation**

This constitutes twisting about an axis (e.g., turning the head from side to side).

V.   Muscle Identification

Each skeletal muscle has two points of attachment: the more fixed parts being the **origin** and the more movable part being the **insertion.** Usually the origin is closer to the median axis of the body, while the insertion is more peripherally located. The two ends of the muscle are attached to different bones, otherwise contraction would not produce movement. When a muscle contracts, the origin usually remains stationary and the insertion is usually moved toward the origin. A muscle can be attached directly to a bone, to another muscle, or to a tendon.

A.   Muscle Location

Listed below are some of the major muscles of the human body. Using the charts available plus the descriptions provided, locate and label each muscle on the following diagram. To better distinguish one muscle from another, color each muscle a different color.

1.   **Deltoideus** — shoulder muscle. Origin: clavicle and scapula. Insertion: upper part of humerus. Action: abduction of humerus.

2.   **Biceps brachii** — muscle of the anterior portion of the upper arm. Origin: scapula. Insertion: radius. Action: flexion of the forearm.

3.   **Sternocleidomastoideus** — long muscle on the side of the neck. Origin: clavicle and sternum. Insertion: mastoid process of the skull. Action: contraction of one causes head to move toward the side; contraction of both simultaneously flexes head forward and downward on the chest.

4.   **Pectoralis major** — triangular muscle in the upper chest. Origin: clavicle, sternum, and cartilages of the 6 upper ribs. Insertion: upper anterior portion of the humerus. Action: adduction and rotation of humerus

5.   **Rectus abdominis** — long flat muscle covering the abdominal region. Origin: pubic bone. Insertion: sternum and cartilages of 5th, 6th, and 7th ribs. Action: flexion of the body at the lumbar region plus compression of the abdominal organs.

6.   **External oblique** — large flat muscle on the side of the body (between rib cage and hip). Origin: external surface of the lower 8 ribs. Insertion: ilium crest and flat band of connective tissue in midline of abdomen. Action: flexion of body plus compression of abdominal organs.

7.   **Sartorius** — a long slender muscle running diagonally across the thigh. Origin: ilium. Insertion: inner surface of the upper part of the tibia. Action: flexion of the leg on thigh and of thigh on the pelvis and rotation of thigh laterally.

8.   **Quadriceps femoris** — very large muscle of the anterior part of the thigh; it consists of 4 parts:

Rectus femoris — located on the anterior part of the thigh.
Vastus lateralis — located on the lateral part of the thigh.
Vastus medialis — located on the medial surface of the thigh.
Vastus intermedium — located underneath the rectus femoris (not visible on the diagram)

Origin: ilium (only on the rectus femoris) and femur (the other 3). Insertion: tendon passing over knee joint to attach to tibia. Action: extend leg (the entire quadriceps) and flex thigh (only rectus femoris).

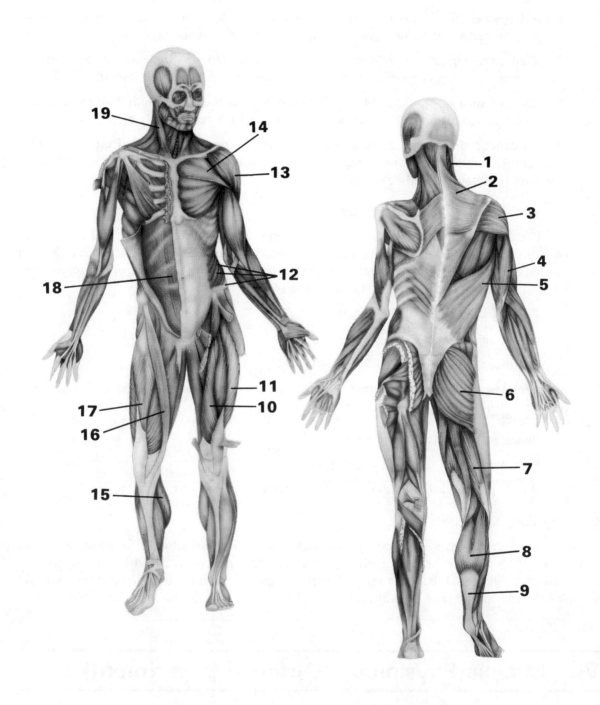

FIGURE 27.1 — HUMAN MUSCULATURE

9. **Trapezius** — triangular muscle in the upper back region. Origin: occipital bone of skull and cervical and thoracic vertebrae. Insertion: scapula and clavicle. Action: adduction of scapula, raising of scapula ("shrugging"), and extension of head backward.

10. **Latissimus dorsi** — back muscle in the lumbar area. Origin: lower 6 thoracic vertebrae, all the lumbar vertebrae, sacrum, and ilium. Insertion: upper part of the humerus. Action: extends, adducts, and rotates arm medially and draws shoulder down and backward.

11. **Triceps brachii** — muscle of the posterior surface of the upper arm. Origin: scapula, and humerus. Insertion: ulna. Action: extends and adducts forearm.

12. **Gluteus maximus** — muscle of the buttocks region. Origin: ilium, sacrum, and coccyx. Insertion: femur. Action: extends, abducts, and rotates the thigh laterally.

13. **Biceps femoris** — muscle of the posterior surface of thigh. Origin: ischium and femur. Insertion: fibula and tibia. Action: flexes leg and rotates laterally after flexion.

14. **Gastrocnemius** — muscle making up the external part of the calf of the leg. Origin: femur. Insertion: calcaneus (heel bone) via tendon of Achilles. Action: extension of foot (points toes) and flexes leg.

B. Action of Muscles During Movement

Perform each of the following actions. When doing so, observe in yourself or your lab partner, the contraction of the muscle involved. By palpitation, feel the muscle harden during contraction. Name the primary muscle responsible for the action, based on the labelled diagrams filled in earlier in this exercise.

Turning head to the side —

Rising on tiptoes —

Bending the knee —

Crossing one leg over the other —

Raising the arm outward —

Bending the forearm at the elbow —

Forcible expiration as in coughing —

Which muscles are usually used for intramuscular injections?

C. Analyzing Movements

As your lab partner goes through the movements indicated by the following diagrams, Figure 27.2, analyze the movement, then draw in the muscle on each figure which causes the movement indicated. Be sure to show where the insertion and the origin are. Also, name the muscle and indicate the type of movement (i.e., adduction, flexion, etc.) produced.

VI. Muscle Physiology (Optional Experiment)

A. Contraction of Cardiac Muscle

Your instructor will pith a frog — a procedure that destroys the brain and spinal cord. The frog's heart will be exposed by dissection.

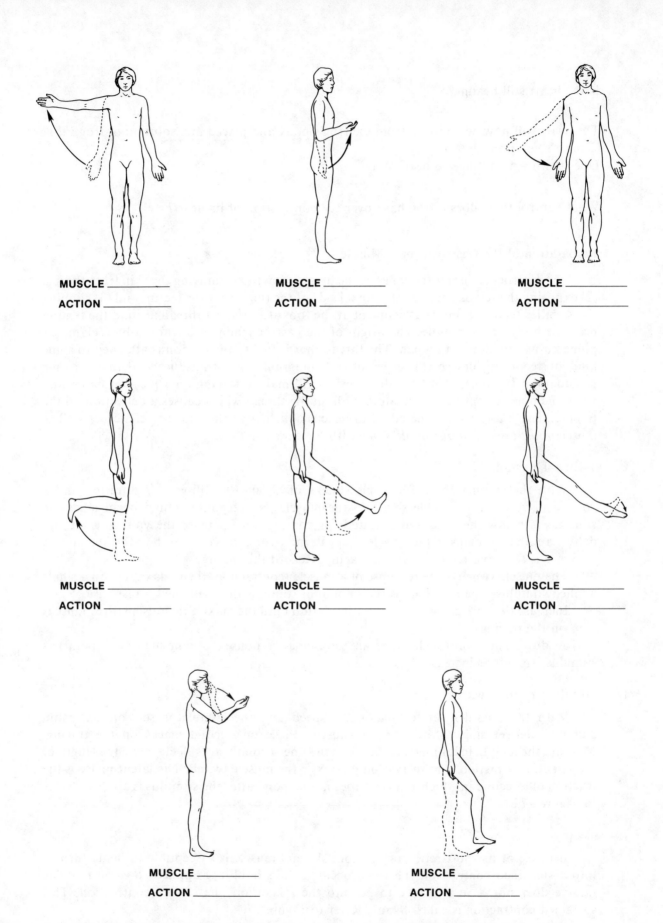

MUSCLE _____
ACTION _____

MUSCLE _____
ACTION _____

MUSCLE _____
ACTION _____

MUSCLE _____
ACTION _____

MUSCLE _____
ACTION _____

MUSCLE _____
ACTION _____

MUSCLE _____
ACTION _____

MUSCLE _____
ACTION _____

FIGURE 27.2 — MUSCLE ACTIONS

Is the heart still beating?

The heart will now be removed from the frog's body and placed in a solution of frog saline (Ringer's solution).
Does the heart continue to beat? Why?

What control then, does a frog have over the contraction of its heart?

B. Preparation of the Gastrocnemius Muscle

The gastrocnemius muscle will now be prepared by: (1) removing the skin from the leg; (2) exposing the sciatic nerve on the dorsal side of the thigh near the femur; and (3) clipping the Achilles tendon from its attachment at the foot after tying a thread around the tendon near the body of the muscle. The origin of the gastrocnemius is anchored by a clamp or pinned down to a dissecting pan. The thread that is tied to the tendon is attached to some kind of recording device (the lever of a kymograph or a transducer of an electronic physiograph machine). The recording device will make a tracing on a piece of paper and when the muscle contracts, it creates a pull on the thread which causes a deflection on the tracing. The muscle is stimulated by an electric current applied to the sciatic nerve. The muscle *must be kept moist at all times* with Ringer's solution!

C. Twitch Threshold

With the stimulator set for single shocks of about 10 milliseconds duration and a voltage of zero, give a single shock to the muscle. No response should occur since the stimulus is below the threshold required for contraction. Increase the voltage a bit and deliver another stimulus to the muscle. Repeat, increasing the voltage a bit at a time, until a response is seen and recorded on the tracing. Record this voltage.

The muscle **twitch** is the response of a muscle (contraction, then relaxation) to a single applied stimulus. The **threshold** is the minimum stimulus that will elicit a response.

Increase the voltage and continue stimulation until the maximum muscular response is seen on the tracing.

How do you account for the increasing muscular response (to some maximum point) as stimulus strength is increased?

D. Duration of the Twitch

With the recorder set for maximum speed and the stimulator set for maximum stimulus, deliver single shocks to the muscle. Place one-second marks on the tracing. Measure the length, in millimeters, between the one-second marks. Determine the length of the **latent, contraction,** and **relaxation phases** of the muscle twitch. The latent phase is the fraction of a second in which a muscle does not respond after the stimulus is applied. Refer to Figure 27.3.

E. Tetany

Because of its short refractory period, skeletal muscle is susceptible to being thrown into a sustained contraction. This will occur if the stimuli are applied so often that the muscle does not have a chance to go into the relaxation phase after contraction. This sustained contraction is called **tetany.** Refer to Figure 27.4.

With the recorder set for medium speed and the stimulator on maximal stimulus, deliver stimuli of 1, 2, 3, 5, 7, 10, and 20 per second in succession to the muscle. Note the fusion of twitches into a tetanic contraction.

Switch off the stimulator, leaving the frequency at 20 stimuli per second. Switch on again and note the smooth contraction which results. Leave on until contraction strength begins to diminish. What is happening to cause this diminished response?

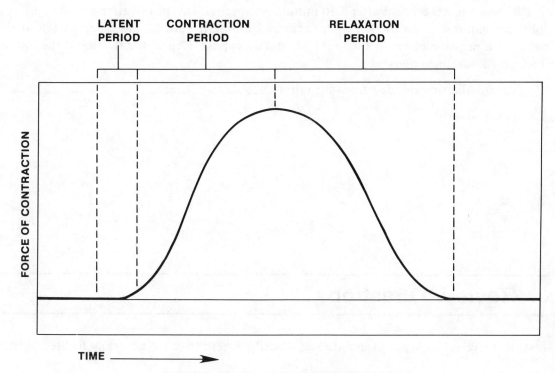

FIGURE 27.3 — THE MUSCLE TWITCH

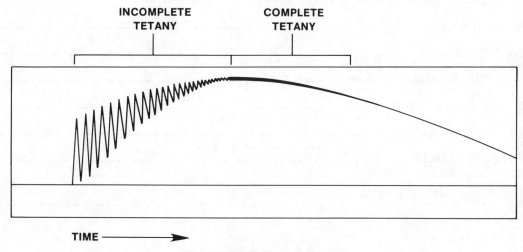

FIGURE 27.4 — TETANY

F.   Fatigue

Repeated contraction of a muscle produces **muscle fatigue.** Fatigue is caused by the accumulation of waste products such as lactic acid, or by a deficiency in oxygen, glucose, and other raw materials.

To experience the sensation of fatigue, lay your left forearm on the table top, palm up. Flex and extend your fingers rapidly at the third joint (where the fingers join the hand) until they are fatigued.

Record the time needed to cause fatigue: _____

Rest the fingers for 10 min., then attach a weight to one finger. Again, record the time

needed to cause fatigue: _____

Repeat the experiment after a 10 minute rest period, except this time put the cuff of a sphygmomanometer on the left upper arm and inflate it until no pulse can be felt at the wrist. At a pressure of about 130 mm it should disappear. Repeat the flexing of the fingers for a given weight as above.

Record the time needed to cause fatigue: _____

Conclusions:

VII. Review Questions

A. List the three major types of muscle and describe where they are generally located.

B. What type of muscle would be found in the following structures?

1. Diaphragm — _____

2. Uterine walls — _____

3. Walls of urinary bladder — _____

4. Ventricle of heart — _____

5. Anal sphincter — _____

6. Biceps brachii — _____

C. Define the following terms:

1. Abduction —

2. Extension —

3. Intercalated disc —

4. Adduction —

5. Involuntary —

D. Which muscle has its origin on the clavicle and scapula, insertion on the upper part of the humerus and has the function of adducting the humerus?

_____

E. Name the four parts of the quadriceps femoris

    1. _____

    2. _____

    3. _____

    4. _____

F. Which muscle is triangular in shape and is located in the upper back? (It's action is the adduction of the scapula, extension of the head backward, and raising the scapula — as in "shrugging.")

_____

G. Which muscle forms the external part of the calf and functions in pointing the toes?

_____

H. Which muscle would be contracting when you cross one leg over the other?

_____

REPRODUCTION
IN ANIMALS

I. Objectives

After the completion of this exercise, the student should be able to do each of the following:

A. Compare the advantages and disadvantages of **asexual** and **sexual** forms of **reproduction.**

B. List three types of asexual reproduction.

C. List the steps involved in **spermatogenesis** and **oogenesis,** and identify the points at which mitosis and meiosis occur in the process.

D. Explain why conjugation in *Paramecium* is a type of sexual reproduction.

E. Describe how **sperm** are formed in the **seminiferous tubules.**

F. Identify the parts of the male and female reproductive systems and describe the function of each part.

G. Describe the changes that occur as a **follicle** develops to maturity in the **ovary** and what happens to the follicle after **ovulation.**

H. Describe the changes in the **endometrium** during the **menstrual cycle.**

I. Explain how the levels of **FSH, LH, estrogen,** and **progesterone** are regulated during the menstrual cycle.

J. Describe the points in the menstrual cycle at which each of the hormones listed above reaches a peak, and describe the effects produced by each.

K. Answer the review questions at the end of the exercise.

II. Introduction

The creation of new cells (or organisms) from previously existing cells (or organisms) has developed along two basic lines: **asexual reproduction** in which no genetic variability exists between parent and offspring and **sexual reproduction** in which genetic variability does exist.

In this lab we will examine various methods of reproduction. The major portion of the lab, however, will be concerned with mammalian reproduction.

III. Asexual Reproduction

Asexual reproduction is common among lower plants and animals and has several advantages: (1) it involves only mitosis; (2) only one parent is necessary, therefore, no complex mating behavior is required; (3) large numbers of offspring can be produced at one time; (4) it is a means of dispersal of offspring; and (5) in a stable environment, it allows for non-variable "copies" of the parent. The major disadvantage is that, because of the genetic stability, there is reduced ability for populations to adapt to changes in the environment. The three main categories of asexual reproduction, which you have already studied, include: **fission** (in protozoa, bacteria, yeast, and some algae); **budding** (as in *Hydra,* sponges, yeast, and some jellyfish); and **fragmentation** or **regeneration** (in flatworms, segmented worms, starfish, and among filamentous forms of algae).

Many organisms reproduce both asexually and sexually. This is particularly true of internal parasites such as malarial organisms and liver flukes, which reproduce asexually during certain parts of their life cycles, thus increasing their numbers significantly, and improving their chances of dispersal. Thus, we find that asexual reproduction is widespread, even persisting, among forms that also reproduce sexually. This is another way of stating that the adaptive advantages of asexual reproduction may outweigh the disadvantages of reduced variation in genetic makeup.

IV. Sexual Reproduction

A. Function and Method

Sexual reproduction is found in the whole range of plant and animal life — from the one-celled level to the multicellular mammalian level. Its primary advantage is that through the process of meiosis and fertilization it allows for genetic variations among organisms, i.e., offspring that are not identical to the parent. This variable is important in that it allows populations to adapt to environmental changes — it is the raw material upon which natural selection acts.

The major disadvantages would be the complex processes of sperm and egg development and the necessity for anatomical and behavioral features to bring sperm and egg together.

In the process of **conjugation** in *Paramecium,* the fusion of nuclei, which is the main feature of sexual reproduction, takes place. The micronuclei of the two organisms divide meiotically. When they unite side by side, they exchange one of their micronuclei. Then the two nuclei fuse in each cell. The two organisms each contain some nuclear material from each other and are different genetically than they were before conjugation. Mitosis follows, resulting in the production of four different offspring from each cell.

Fertilization in animals involves the fusion of nuclei in specialized gametes, one large and nonmotile, the other small and motile. In some organisms, such as hydra, earthworms, and tapeworms, one individual may produce both sperm and eggs, although they usually do not fertilize their own eggs. In other cases (among invertebrates especially, e.g., aphids, and

302

bees), the unfertilized egg may develop into an adult; a process called **parthenogenesis.** In those organisms in which there are separate sexes, producing unique gametes, there are numerous variations on the means of fertilization. In some cases (e.g., frogs) the gametes are shed into the water and fertilization occurs there. In other cases, development of special structures is necessary for fertilization to take place internally. In the rest of this lab we will deal with sexual reproduction in the latter group of animals with an emphasis on mammals.

B.   Gametogenesis

      Gametogenesis is the process involving the development of gametes or sex cells. Gametogenesis in the male is called **spermatogenesis** (development of sperm); in the female it is called **oogenesis** (development of the ovum). Both meiosis and mitosis are involved in gametogenesis. During embryonic development certain cells become different from the other body cells; these are the primordial germ cells which will give rise to the sex cells. Each primordial germ cell undergoes many mitotic divisions. producing a large number of cells. These cells are diploid, therefore it is necessary for them to undergo meiosis in order to develop into functional gametes.

      Figure 28.1 shows the general scheme of spermatogenesis and oogenesis. Be sure that you understand why each functional sex cell contains only one-half the chromosomes that are in the primordial germ cells.

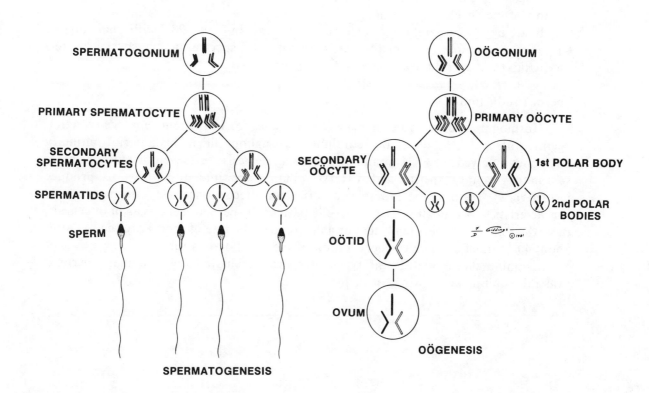

FIGURE 28.1 — SPERMATOGENESIS AND OOGENESIS

303

C. Male Reproductive System

1. Testis

Under low power, examine a prepared slide of monkey testis. Note the round or oval cross sections of the **seminiferous tubules** in which the sperms are produced. Tubules are held together by loose connective tissue, which contains blood vessels, nerves, and **interstitial cells.** The latter are compact masses in the spaces between adjacent tubules, not particularly distinct from connective tissue. Such interstitial cells are endocrine; they are the main producers of the male sex hormones, or **androgens.**

Study a few seminiferous tubules under high power. The wall of each tubule contains an outer **basement membrane** which forms the boundary of several layers of stratified epithelial cells. Most of the cells of this epithelium are **spermatogonia,** the precursors of sperms.

The spermatogonia represent the diploid **generative epithelium** of the testis. The cells divide mitotically and mitotic stages may actually be visible.

As a result of the continuing divisions, some of the daughter cells, called **spermatocytes,** become displaced toward the **lumen** (central cavity) of the tubule. Spermatocytes then undergo meiosis and become **spermatids,** haploid spherical cells. The spermatids differentiate into mature sperms, with their tails projecting into the lumen of the tubule. Subsequently, the sperms come to lie free in the lumen. New sperms continuously form, eventually displacing the older ones, and the latter are then gradually pushed out of the testicular tubules into the sperm duct.

Scattered among the developing sex cells are relatively large cells attached to the basement membrane but extending inward to the lumen of the seminiferous tubule. These cells are known as **Sertoli cells** and are believed to function in the nutritive support of the developing sperm.

Examine the slide and try to correlate the diagram in Figure 28.2 with what you see. On the diagram, note the interstitial cells, Sertoli cells, spermatogonia, spermatocytes, spermatids, and sperm (developing and complete).

2. Sperm Production

Hormonal production in the male is not cyclical as in the female. Once secretion begins with puberty, it usually continues throughout life. The anterior pituitary hormones involved are **FSH** (stimulating spermatogenesis, i.e., development of primary spermatocytes into sperm) and **LH** (stimulating the interstitial cells to produce **testosterone**). Testosterone promotes a variety of effects concerned with male sexual characteristics. Although there is no cyclic pattern in spermatogenesis as in oogenesis, there does appear to be a tendency for a decrease in the rate of sperm production after about 40 years of age due to degeneration of the seminiferous tubules.

Examine slides of different types of sperm (including living human sperm if available) and make drawings of them.

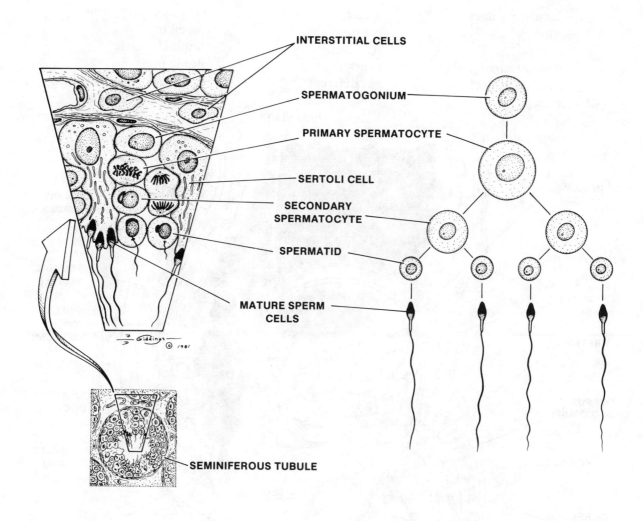

INTERSTITIAL CELLS

SPERMATOGONIUM

PRIMARY SPERMATOCYTE

SERTOLI CELL

SECONDARY
SPERMATOCYTE

SPERMATID

MATURE SPERM
CELLS

SEMINIFEROUS TUBULE

FIGURE 28.2 — CROSS-SECTION OF A MAMMALIAN SEMINIFEROUS TUBULE

3. Anatomy

The main function of the reproductive system is to insure continuation of species. The organs of this system are designed so that the sex cells (eggs and sperm) are brought together in the reproductive system of the female, so that fertilization may occur internally.

The male reproductive system consists of **testes, epididymides** (sing. - **epididymis**), **vasa deferentia** (sing. - **vas deferens**), and their associated glands (**prostate, bulbourethral glands,** and **seminal vesicles**), and **external genitalia** (the **penis,** and **scrotum**).

Using Figure 28.3 as a guide, locate the following structures on the model of the male pelvis:

ejaculatory duct	ureter	corpus cavernosum
colon	prostate gland	foreskin
urethra	scrotum	seminal vesicle
inguinal canal	penis	pubic bone
glans penis	vas deferens	ampulla of vas deferens
testis	urinary bladder	corpus spongiosum
epididymis	bulbourethral gland	

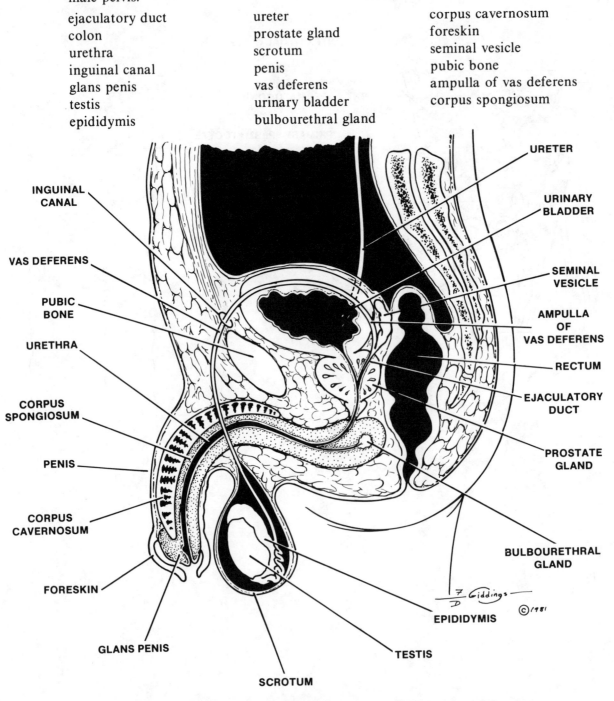

FIGURE 28.3 — MALE REPRODUCTIVE SYSTEM

D. Female Reproductive System

1. **Anatomy**

The structures of the female reproductive system are homologous to those of the male reproductive system. They include **ovaries, oviducts, uterus, vagina,** and **external genitalia (clitoris, labial folds,** and **mons pubis).** The **mammary glands** of the female are also included in the reproductive system, but actually they should be included in the study of the skin, since they are modified sweat glands.

Find all of the parts listed below on the model of the female pelvis using Figure 28.4 as a guide.

ovary	uterus	labium major
oviduct	rectum	pubic bone
cervix	vagina	labium minor
urinary bladder	mons pubis	urethra
clitoris	ligament	

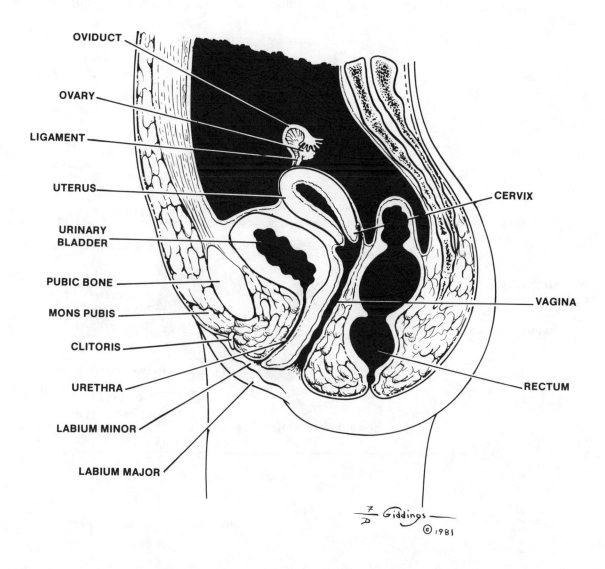

FIGURE 28.4 — FEMALE REPRODUCTIVE SYSTEM

307

2. Ovary

As you read the rest of this exercise, refer frequently to Figure 28.5. In a section through a cat ovary, the outer tissues form the cortex, a layer not marked off sharply from the core tissues, or medulla. The surface layer of the cortex represents the **generative epithelium** from which strands of cells are budded off.

Below the cortical surface, note the presence of a distinct zone containing **developing follicles.** Each such follicle contains an **oocyte,** or immature egg, substantially larger than other cells in the vicinity. Directly surrounding the oocyte is a layer of more or less cuboidal **follicle cells.** At this stage the oocytes have usually completed the first meiotic division; the second division will occur after the egg leaves the ovary. In humans, the second division will take place only if the egg is penetrated by a sperm cell.

As the follicle (oocyte plus follicle cells) matures and grows, it ultimately comes to occupy the entire width of the ovarian cortex. Follicle cells divide (mitotically) and produce the female sex hormones, or **estrogens.** Follicle growth is accompanied also by the appearance of an enlarging fluid-filled cavity. In one region of the follicular wall a **germ hill** exists, which supports and surrounds the oocyte. When ovulation occurs, the oocyte escapes through ruptures in the follicular wall and the cortical tissue of the ovary.

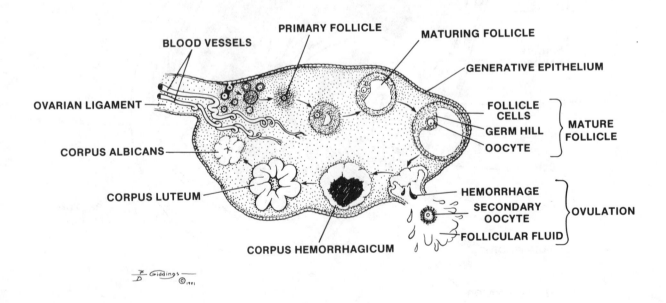

FIGURE 28.5 — OVARY SHOWING FOLLICLE DEVELOPMENT, OVULATION, AND FORMATION OF A CORPUS LUTEUM

Examine a prepared slide showing a mammalian ovary with a mature follicle. Make a diagram of it below, labelling the mature follicle, oocyte, follicle cells, and primary follicle.

E. **Menstrual Cycle**

1. Phenomena Occurring During the Cycle

The menstrual cycle involves cyclic changes occurring in the anterior lobe of the pituitary, ovaries, and uterus of the female. These changes deal with the preparation of the egg for, and its release at, the time of ovulation. In addition, the uterus undergoes changes in preparation to receive the developing embryo.

The textbook "average" for the length of the cycle is 28 days, with ovulation generally occurring about 14 days *before* the onset of the next menstrual flow period.

The steps involved in this cycle are (refer to Figures 28.6 through 28.8):

(a) The high concentration of **FSH (Follicle Stimulating Hormone** from the anterior pituitary) in the bloodstream during the beginning of the cycle stimulates mitotic division in some of the follicles during the first half of the cycle.

(b) The follicle cells begin producing **estrogen.** When the estrogen level gets above a certain threshold, it inhibits the production and release of FSH from the pituitary. This hormone is responsible for causing an increase in the thickness of the endometrium (the inner lining of the uterus where the fertilized ovum would implant). Later in the cycle, when the estrogen level drops, the level of FSH will begin to increase. This FSH-estrogen interaction is an example of a **negative feedback mechanism,** in which one substance stimulates the production of a second substance and the second substance inhibits the first.

(c) Estrogen in the bloodstream stimulates the secretion of **LH (Luteinizing Hormone)** from the anterior pituitary. Therefore, as the estrogen level increases, the increasing level of LH stimulates further development of the follicle and causes **ovulation** (the release of the secondary oocyte from the mature follicle) at around day 14 of the cycle.

(d) At the time of ovulation, some of the capillaries that had been nourishing the developing follicle rupture and release blood into the cavity of the follicle. What remains of the follicle together with its blood clot is known as the **corpus hemorrhagicum.** The clot will be removed by white blood cells, and with the removal of the clot and the thickening of the cell layers, the corpus hemorrhagicum becomes the **corpus luteum,** capable of producing progesterone as well as small amounts of estrogen. While the secondary oocyte moves down the oviduct, the progesterone stimulates the endometrium to undergo preparations for the implantation of an embryo, should fertilization occur. Once the level of progesterone in the bloodstream increases, however, another negative feedback mechanism goes into effect, in which the secretion of LH is inhibited.

(e)   In the non-pregnant woman, the corpus luteum lasts for about 10-12 days and begins to degenerate on about day 20 or 21, due to the drop in the level of LH in the blood. The corpus luteum degenerates to form the **corpus albicans,** which gradually shrinks in size until only a microscopic scar remains on the ovary to indicate its location. This degeneration causes a decrease in the level of progesterone and estrogen (from the corpus luteum) arriving at the uterus via the bloodstream. As a result, the arteries supplying blood to the endometrium contract, causing the endometrial tissue to start degeneration and sloughing off (**menstruation**) due to an inadequate blood supply.

(f)   As soon as the level of estrogen drops due to the degeneration of the corpus luteum, the level of FSH increases as a result of the feedback mechanism mentioned in (a). FSH then begins the cycle over again.

(g)   If pregnancy does occur, steps (e) and (f) are eliminated. The egg is usually fertilized in the upper 1/3 of the oviduct. The embryo, specifically the trophoblast cells of the blastocyst, and later the developing chorion, produces **HCG (human chorionic gonadotropin).** This hormone functions to maintain the corpus luteum and thus maintain the level of progesterone necessary to maintain the endometrium and sustain pregnancy.

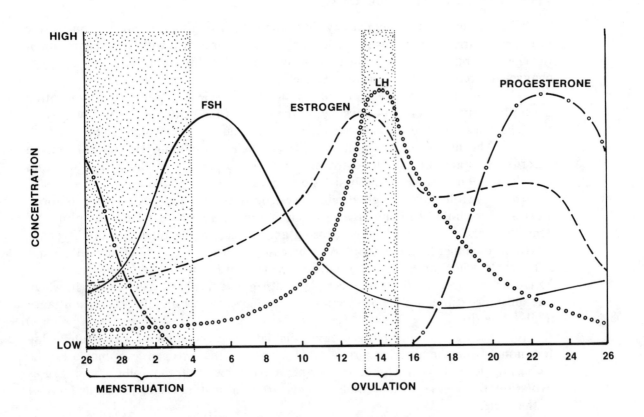

FIGURE 28.6 — PLASMA HORMONE LEVELS DURING MENSTRUAL CYCLE

2. Study demonstration slides of ovaries containing corpora hemorrhagica, corpora lutea, and corpora albicans, diagramming each in the space below.

3. Examine demonstration slides of the endometrium at different stages of thickening. Make sketches in the space provided and compare these to Figure 28.7.

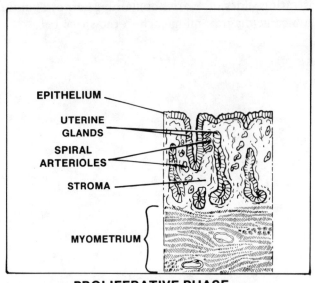

PROLIFERATIVE PHASE

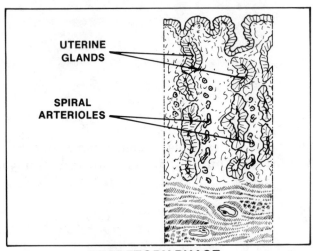

SECRETORY PHASE

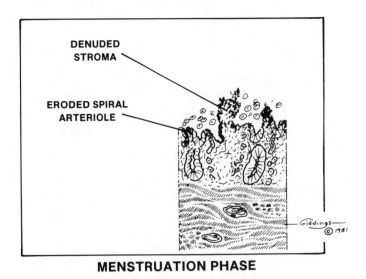

MENSTRUATION PHASE

FIGURE 28.7 — DIAGRAMMATIC REPRESENTATION OF THE UTERINE ENDOMETRIUM

In Figure 28.8 showing changes in the ovary and endometrium during the menstrual cycle, identify:

follicle
corpus luteum
menstrual flow days
proliferative phase (estrogen stimulated thickening of the endometrium)
secretory phase (progesterone stimulated glandular development of the endo-
        metrium)
corpus albicans
ovulation time
corpus hemorrhagicum
secondary oocyte

For the four hormones important in this cycle, indicate their sources, peak levels during the cycle, and their functions.

V. Contraception

Various types of methods and devices are currently used to prevent pregnancy. Each of them varies in its effectiveness in preventing conception and in safety for the individual using it. Among these methods and devices are: **oral contraceptives, intrauterine devices (IUD's), sterilization,** mechanical barriers (**condom, diaphragm**), **vaginal spermicides,** and the **rhythm method.**

Observe the film in lab entitled "Contraception" which explains these methods fully and discusses their effectiveness.

VI. Review Questions

A.   If a menstrual cycle is regularly 34 days, on about what day would you expect ovulation to

occur? Why? _____

_____

B.   If it is correct to say that one hormone is primarily responsible for the onset of menstruation,

which one might it be? Why? _____

_____

C.   In what organs of the reproductive system (male and female) would there be a high rate of

mitosis? _____

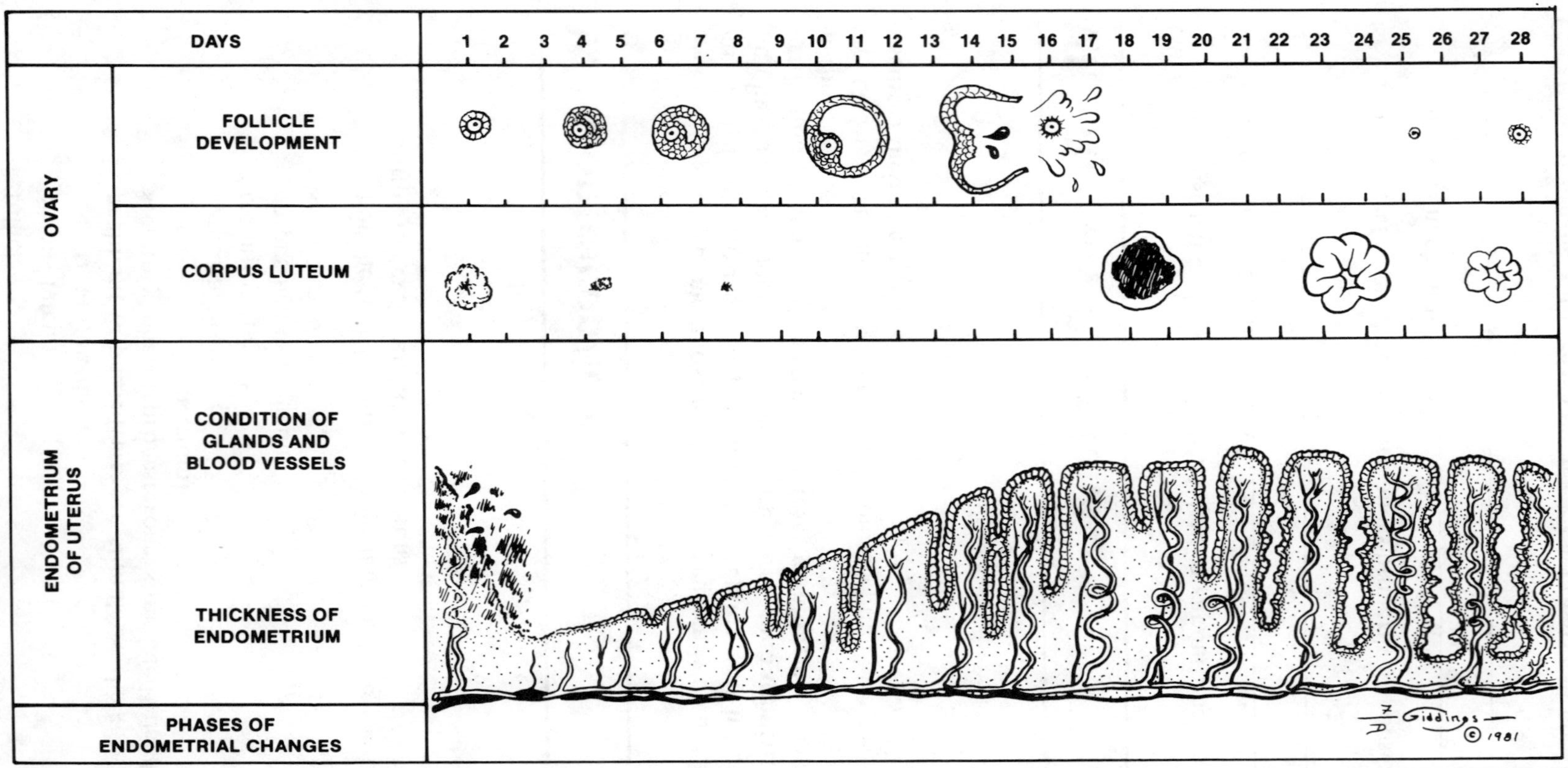

HORMONES:	SOURCE	FUNCTION	WHEN PEAK LEVELS OCCUR
LH			
FSH			
ESTROGEN			
PROGESTERONE			

FIGURE 28.8 — DIAGRAMMATIC REPRESENTATION OF INTERRELATIONSHIPS OF REPRODUCTIVE HORMONES AND THE OVARIAN AND UTERINE CYCLES

D. Which structures contribute to the formation of semen? _____ 315

_____

E. Where is the corpus luteum located during its functional life? _____

F. What does the corpus luteum develop from? _____

G. What happens to the corpus luteum when it is no longer functional? _____

_____

H. Where is an IUD placed? _____

I. What structure in the female would have the same embryological origin as the penis in the male? _____

J. What is the function of seminal vesicles, prostate, and bulbourethral glands? _____

_____

K. List in sequence the structures which sperm must travel through in the male reproductive system during an ejaculation. _____

_____

_____

COMPARATIVE EMBRYOLOGY

I. Objectives

After the completion of this exercise, the student should be able to do each of the following:

A. Identify the major stages of starfish development under the microscope.

B. Define **fertilization, cleavage, blastula, gastrula,** and **morphogenesis.**

C. Identify the major stages of frog and chick development under the microscope.

D. List the differences in the four types of embryonic development discussed.

E. Explain the function of each of the four extraembryonic membranes.

F. List the three germ layers and name at least two structures derived from each.

G. Answer the review questions at the end of this exercise.

II. Introduction to Embryology

In the process of sexual reproduction in animals, two gametes — sperm and egg — fuse to make one new cell. These two gametes are usually donated by two different parents. The new zygote resulting from this fusion contains material from both parents and is "new" in the sense that it contains new potentialities resulting from the mixture of the genetic material from the two parents. Embryology is the study of the development of the zygote, the product of sexual reproduction.

The zygote develops as a result of three kinds of activity: **mitosis** and subsequent growth, **differentiation of cells,** and **movement of cells.** Thus, as a result of embryonic activity, the one-celled fertilized egg (zygote) changes or develops into the adult form, eventually. In observing the embryonic development of any animal, we can divide it into the following stages:

A. **Fertilization** — the fusion of nuclei and other events associated with the union of sperm and egg. The result is the zygote.

B. **Cleavage** — a series of mitotic divisions undergone by the zygote. No growth in size of the entire structure occurs at this time. The actual pattern of cleavage — whether the whole zygote (**holoblastic cleavage**) or only a part of it divides (**meroblastic cleavage**) — varies from one species of animal to another, depending on the amount of yolk in the egg and its distribution. The end result of cleavage is the formation of a hollow ball of cells known as a **blastula,** the central cavity of which is known as the **blastocoel.**

C. **Gastrulation** — the migration of cells of the blastula resulting in the formation of a new cavity known as the **gastrocoel** (archenteron, or "primitive gut"). Further development leads to the formation of three distinct **germ layers** (embryonic cell layers) known as the **ectoderm, mesoderm,** and **endoderm,** from which all the organs of the new organism develop. At this stage, when the three germ layers are present, the structure is technically termed an embryo.

   The development of internal shape characteristics of the animal, known as **morpho-genesis**, results from the shaping of the germ layers due to differential growth, movement, and association of cells in the germ layers.

D. **Neurulation** — the development of the notochord, neural tube, and coelom. This only occurs in chordates.

E. **Organogenesis** — the differentiation and association of cells to form organ systems.

   In this exercise, we will make a brief survey of four different animals: starfish, frog, chick, and human. This survey should, first of all, illustrate the basic similarities in the embryonic development of these entirely different animals. Secondly, even though there are similarities, there are also variations, based primarily on the amount of yolk present in the cytoplasm of the egg.

III. Starfish Development

Obtain one slide of the starfish development. On this slide, you will find all of the stages indicated in Figures 29.1-29.7. The stages on the slides are whole mounts, not sections. Identify each of the stages indicated on the slide.

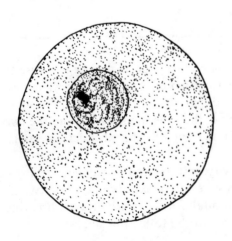

A. **Unfertilized egg:** In this nearly spherical cell, a large nucleus and a nucleolus are clearly visible. A small amount of yolk (stored food) is present in the form of many small particles. Are these yolk particles present in any particular area of the cytoplasm, or are they scattered about?

**FIGURE 29.1
UNFERTILIZED EGG**

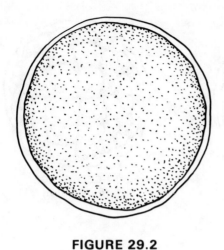

B. **Zygote** (fertilized egg): This is a single celled structure, very much like the unfertilized egg in appearance. In contrast to the unfertilized egg, the
zygote's nucleus is inconspicuous. Would this structure be 1N or 2N?

**FIGURE 29.2
ZYGOTE**

C. **Early Cleavage:** The 2, 4, and 8-celled stages are included in early cleavage. Find an
example of each of these and note that the cells remain attached to each other. Is there any
growth in size?

How does the cell size in each of these stages compare with the size of the zygote?

Are all the cells within a single stage the same size?

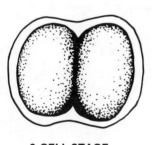
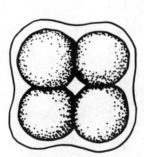
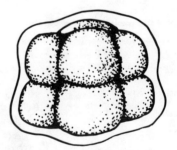

2-CELL STAGE   **4-CELL STAGE**   **8-CELL STAGE**

FIGURE 29.3 — EARLY CLEAVAGE

D. **Later Cleavage:** The 16, 32, and 64-celled stages are included in later cleavage. The 64-celled
stage is hollow and is called the **blastula.**

What happened to the individual cell size?

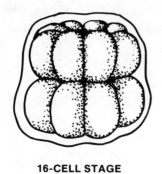
**16-CELL STAGE
(MORULA)**

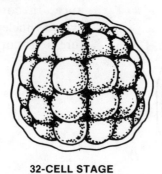
**32-CELL STAGE
(EARLY BLASTULA)**

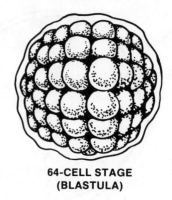
**64-CELL STAGE
(BLASTULA)**

FIGURE 29.4 — LATER CLEAVAGE

E. **Blastula:** As cell division continues, the increasing number of cells become arranged around an enlarging central cavity known as the **blastocoel.** In the starfish, the walls of the blastula are usually one cell layer thick. Why would you expect your specimen to appear dark around the edges and light in the middle?

   Can you see any differences among the cells?

   How does the size of the blastula compare to earlier stages?

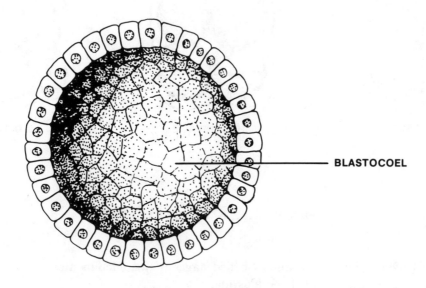

BLASTOCOEL

FIGURE 29.5 — LATE BLASTULA CROSS-SECTION

320

F. **Gastrula:** Shortly after the formation of the blastula, a small depression begins to appear at one end of the embryo; the appearance of this depression marks the beginning of gastrulation. As gastrulation proceeds, the depression invaginates (folds inward) more and more. Which of the three basic embryonic activities would this be?

Find the embryos at various levels of gastrulation on your slide. The latest stages on your slide are those in which the inner end of the invagination is beginning to expand.

As a result of gastrulation in the starfish, the embryo produces two primary cell layers: an outer **ectoderm,** and an inner **endoderm.** The third layer, the **mesoderm,** develops later, between these two. Gastrulation, then, eventually results in an embryo with 3 primary germ layers, a mere remnant of the blastocoel, and a new cavity known as the **gastrocoel.** The gastrocoel is continuous with the outside through the **blastopore** and which will become the cavity of the digestive tract. In the deuterostomes, the blastopore will eventually become the anus.

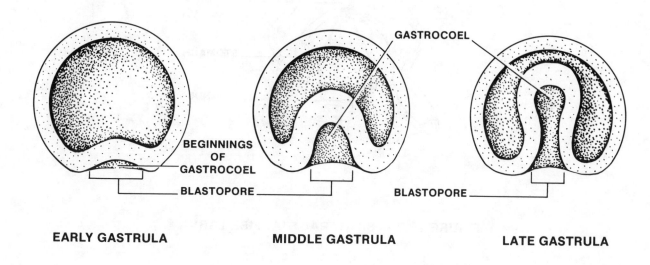

FIGURE 29.6 — GASTRULATION

G. **Larval Stage:**

The gastrula stage is reached within a day or two after fertilization. Within another day or two, this stage undergoes some alterations to give rise to the **larval stage,** which is free-swimming. During a period of from several weeks to several months, the larva grows. After this time, it settles to the bottom and becomes a small starfish. Refer to Figure 29.7.

IV. Frog Development

Examine the charts, models, and whole specimens available in lab. Obtain slides of sectioned embryonic stages. Using Figures 29.8 through 29.15, locate the structures printed in boldface in this part of the exercise.

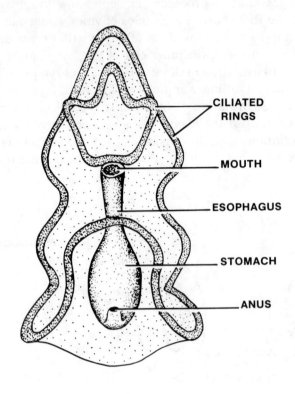

FIGURE 29.7 — BILATERAL STARFISH LARVA

A. **Unfertilized Egg:** The egg consists of two portions; a darkly pigmented portion, the **animal pole,** and a lightly colored, yolk-filled portion, the **vegetal pole.** In nature, when the egg is released into the water, the **gelatinous covering** produced by the oviduct absorbs water and swells, causing the eggs to be equidistant from each other. What is the importance of this swelling?

B. **Zygote:** Sperm have to penetrate the eggs before swelling of the gelatinous covering takes place. Once fertilization has occurred, the cell rotates in such a way that the heavier portion of the cell, the yolk-filled vegetal pole, is downward.

Suppose you had a dish of fertilized frog eggs; how would you know if they were all fertilized (i.e., what color would be facing upward)?

Another indication that fertilization has occurred is the appearance of a pigmented area **(gray crescent)** between the yolk-filled and black portions.

322

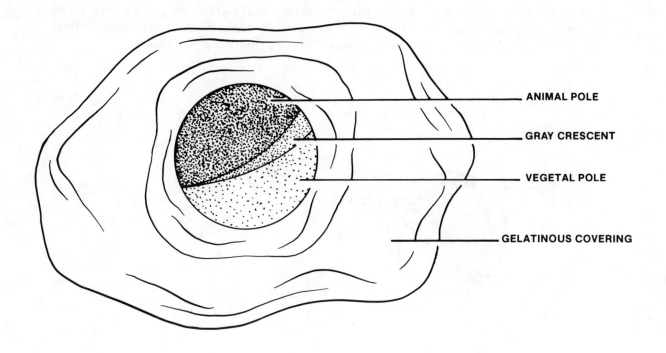

Figure 29.8 — FROG ZYGOTE

C. **Cleavage:** The beginning of the first cleavage is marked by the appearance of a groove on the animal pole end of the egg; this **cleavage furrow** gradually extends toward the opposite side of the zygote, dividing it into two cells. The next cleavage occurs at right angles to the first and produces the **4-cell stage.** The third cleavage occurs parallel to, but a little above the equator of, the developing embryo. What would you suspect was the cause of this unequal division?

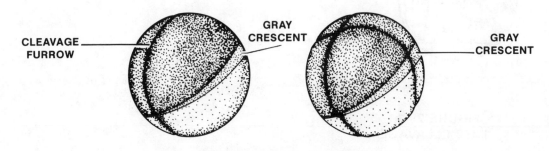

FIGURE 29.9 — EARLY CLEAVAGE

In which region of the developing embryo would you expect a more rapid rate of cell division?

What effect would this have on cell size in the two regions of the developing embryo?

323

Below are two diagrams of early and late cleavage. Note that the jelly layer is not shown in these and subsequent stages. In the space beside each of the drawings, make a diagram of the section through these stages as seen in the prepared slides. Label the **animal pole, vegetal pole,** and **cleavage grooves** on your drawings. Can you find the yolk particles?

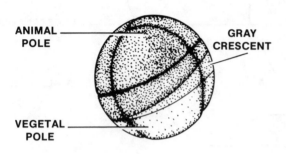

FIGURE 29.10
THIRD CLEAVAGE

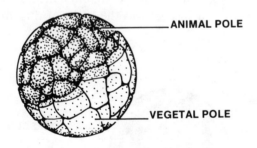

FIGURE 29.11
LATE CLEAVAGE

Give two ways in which the animal pole can be distinguished from the vegetal pole.

Is the **pigment** localized in any particular region of the cell cytoplasm or is it distributed equally throughout?

What do you think is the function of the **membrane** (blue-purple color) around the entire structure on the slides?

D. **Blastula:** The stage in which the **blastocoel** (internal cavity) is formed is called the blastula. Notice that, instead of being centrally located as in the starfish, the blastocoel is off center toward one pole of the developing embryo.

In which hemisphere is it located? Why?

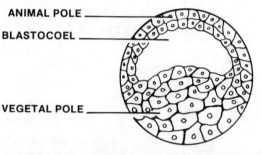

ANIMAL POLE
BLASTOCOEL
VEGETAL POLE

LONGITUDINAL SECTION, BLASTULA   **INTACT LATE BLASTULA**

FIGURE 29.12 — FROG BLASTULA

E. **Gastrula:** (**Yolk Plug** Stage): The large amount of yolk in the frog egg prevents the type of invagination at one end of the blastula which is seen in the starfish. Instead, the more rapidly dividing cells from the animal pole grow downward *over* the yolk-filled cells, gradually enclosing them. **Gastrulation** begins with the pushing inward (or **invagination**) of these cells. This forms a crescent-shaped "line" on the surface known as the **dorsal lip.** The opening marked by the dorsal lip is known as the **blastopore.** This marks the posterior end of the embryo. In the prepared slide of the early gastrula, you can actually see a depression forming here as the surface cells move inward. The depression gradually enlarges to form the **gastrocoel,** as seen in the slide of the late gastrula. Note the difference between this gastrocoel and the one in the starfish.

What will the gastrocoel give rise to?

With continued cell division and the migration of cells inward, the animal pole cells encircle, and eventually line, the entire gastrocoel except for a small circular plug of yolk cells within the blastopore opening, the **yolk plug.** Eventually, even the yolk plug disappears as it is covered by the migrating cells. As all of this is occurring, the three developing **germ layers** are becoming arranged in such a way that there is an outside layer of cells (**ectoderm**), an inner lining of the gastrocoel (**endoderm**), and a layer between the two (**mesoderm**).

Can you detect any visible differences between the cells of these three layers?

Is there any difference in the overall shape of the structure?

325

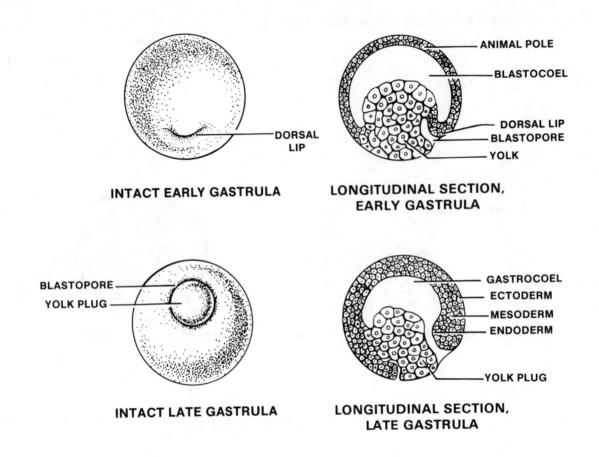

Labels on figure:
- INTACT EARLY GASTRULA — DORSAL LIP
- LONGITUDINAL SECTION, EARLY GASTRULA — ANIMAL POLE, BLASTOCOEL, DORSAL LIP, BLASTOPORE, YOLK
- INTACT LATE GASTRULA — BLASTOPORE, YOLK PLUG
- LONGITUDINAL SECTION, LATE GASTRULA — GASTROCOEL, ECTODERM, MESODERM, ENDODERM, YOLK PLUG

FIGURE 29.13 — FROG GASTRULATION

F. **Neurula:** Near the end of gastrulation, the ectodermal cells in the mid-dorsal region of the embryo thicken to form a flattened area on the surface known as the **neural plate.** The sides of the neural plate, the **neural folds,** gradually fold upward forming a depression, the **neural groove,** between them. Eventually the folds fuse, forming a closed tube, the **neural tube,** which will develop into the brain and spinal cord.

In the midline of the mesodermal layer, the cells develop into a cylindrical rod, the **notochord.**

Although the notochord is a rod extending along the length of the animal, how would you expect it to appear in your slide?

What is the cavity in the area below the notochord?

What is the function of the large cells that may be (depending upon where the animal was sectioned) located in this region?

326

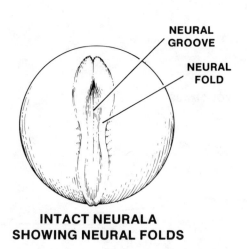

NEURAL
GROOVE

NEURAL
FOLD

**INTACT NEURALA
SHOWING NEURAL FOLDS**

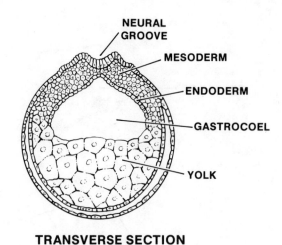

NEURAL
GROOVE

MESODERM

ENDODERM

GASTROCOEL

YOLK

**TRANSVERSE SECTION
THROUGH NEURAL FOLDS**

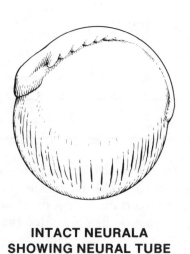

**INTACT NEURALA
SHOWING NEURAL TUBE**

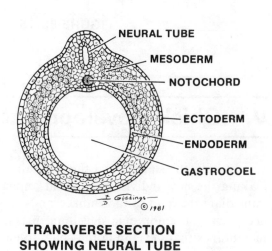

NEURAL TUBE

MESODERM

NOTOCHORD

ECTODERM

ENDODERM

GASTROCOEL

**TRANSVERSE SECTION
SHOWING NEURAL TUBE**

FIGURE 29.14 — NEURULATION

G. **Larval Stage:** With the formation of the neural tube, the embryo elongates. The anterior end becomes slightly enlarged, forming the **head.** A **tail** develops from the posterior end, and **gills** develop for gas exchange. The mouth opens so that feeding can occur. Over a period of growth of a few months, this **tadpole** will metamorphose into an adult frog capable of survival both on land and in water.

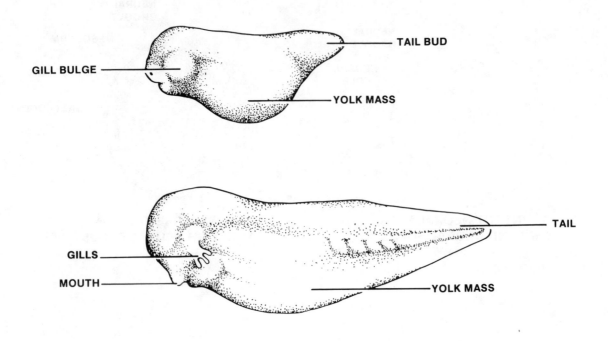

GILL BULGE

TAIL BUD

YOLK MASS

GILLS

MOUTH

TAIL

YOLK MASS

FIGURE 29.15 — FROG LARVAL STAGES

V.   Chick Development

In both the starfish and the frog, all of the cells derived from the fertilized egg are used in making the new individual. In land animals, however, a number of cells are not used in the immediate makeup of the embryo body; instead, they give rise to certain temporary structures necessary for embryonic development on land. The diagram below demonstrates the four membranes that develop in the chick to ensure survival:

Amnion — contains water to shield the embryo.

Yolk sac — contains yolk as a food source.

Allantois — stores waste materials produced by the developing embryo plus transporting respiratory gases between the embryo and its environment.

Chorion — also functions in exchange of gases.

Thus, the egg of the chick is a "self-contained" capsule simulating the watery environment of the lower vertebrates and yet showing remarkable adaptations to land development.

A.   **Unfertilized egg:** Examine the egg on demonstration in lab. Figure 29.16 shows the internal structure of the chicken egg. Notice that in this case the everyday term of "egg" refers to more than simply the egg cell, **ovum.** The circular yellow mass that we would refer to as "yolk" as we are sitting at the breakfast table is actually the ovum. The yolk of the egg is in fact the single egg cell. It is composed of a large amount of **yolk granules** plus a small amount of yolk-free cytoplasm, the **germinal disc,** on the surface of the yolk. It is the germinal disc that contains the nucleus to be fertilized and that will undergo cell division to form the embryo. If the egg becomes fertilized, the germinal disc is referred to as the **blastoderm.** The yolk supplies nourishment to the developing embryo.

After the embryo has been ovulated, it will pass down the oviduct acquiring a number of other structures designed to aid survival in the land environment. Immediately around the ovum is the egg "white"; this is known as **albumen,** which functions as a water reservoir for the young embryo and as a food source for later development. Attached to the yolk and extending out into the albumen are dense cordlike structures, the **chalazae,** that serve to suspend the yolk in the albumen. In another region of the oviduct, two thin shell membranes are deposited around the albumen; these function in decreasing water loss. The next section of the oviduct produces the **shell** and molds the egg into its customary shape. The hard protective shell is porous, allowing for gas exchange.

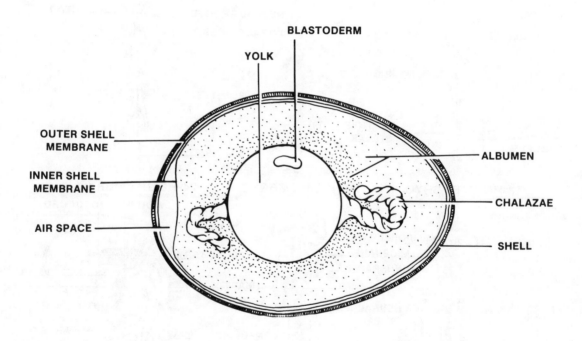

FIGURE 29.16 — INTERNAL STRUCTURES OF HEN'S EGG

B. **Early Development:** By the time the egg is laid, the stages of development through gastrulation have usually occurred, and the embryo is undergoing neurulation. Notice that only the blastoderm undergoes cleavage. Note also that, because of the tremendous amount of yolk, the blastula is not a "hollow ball of cells"; instead, it is a hollowed, flattened disc on the yolk surface. A third difference is with regard to gastrulation. Notice that the blastopore (the site where surface cells migrate inward to establish the 3 germ layers) is not a circular opening as was in the case of the starfish and the frog. In the chick, the blastopore is elongate and is referred to as a **primitive streak.**

C. *Later Development:* Obtain plastomounts of different stages of chick development and examine them under a stereomicroscope for the structures in Figure 29.17. Notice particularly the characteristic curvature of the embryo body as it is being separated from the underlying yolk. The circulatory system within the embryo is quite extensively developed; the heart would definitely be beating if this were alive. Why do you suppose there is such an extensive blood vessel network outside the embryo body? The 72-hour embryo has two extreme bends which make it curve back on itself. Identify the **anterior limb bud** which will give rise to the wing.

Notice the bump-like structures on either side of the spinal cord (middorsal region). These structures are the **somites,** blocks of mesodermal cells that will later give rise to muscles and vertebrae. They are also used as a means of determining how old the embryo is. How many pairs are there?

In looking at the embryo, can you determine if there is a difference in the rate of development in comparing the anterior and posterior regions of the body? Explain your answer.

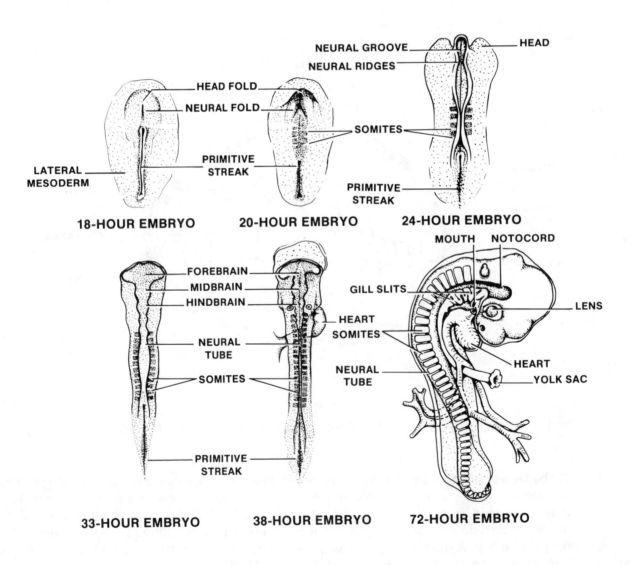

FIGURE 29.17 — STAGES IN THE DEVELOPMENT OF THE CHICK

D. Chick development continues for 21 days, until hatching. Note that this animal does not go through a larval stage as did the starfish and frog.

VI. Extraembryonic Membranes of the Chicken and Mammals

Human development parallels that of the chick even though the egg of humans does not contain an overabundance of yolk. Why? _____

_____

_____

The reptiles were the first to lay eggs on land. Their eggs contained extraembryonic membranes by which the embryo carried out gas exchange, excretion of wastes, and consumption of stored food (yolk). These same membranes develop in the human, but are put to different uses since the human develops internally. Figure 29.18 shows the membranes in a chick egg and compares them to those in the human.

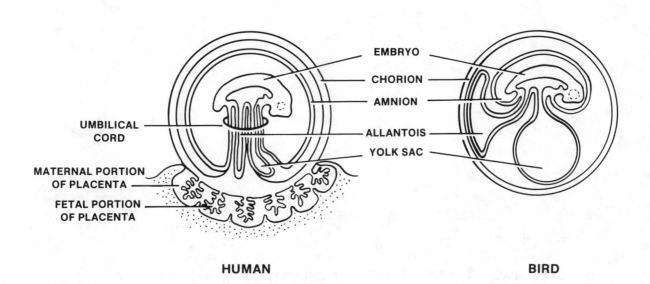

HUMAN   BIRD

FIGURE 29.18
COMPARISON OF THE EXTRAEMBRYONIC MEMBRANES IN THE HUMAN AND CHICK

VII. Human Development

Referring to Figure 29.19, note that fertilization normally occurs in the upper 1/3 of the oviduct, and early development of the embryo occurs during its movement down the oviduct. By the time the developing embryo reaches the uterine cavity (5-7 days after ovulation), it is in the **blastocyst** stage.

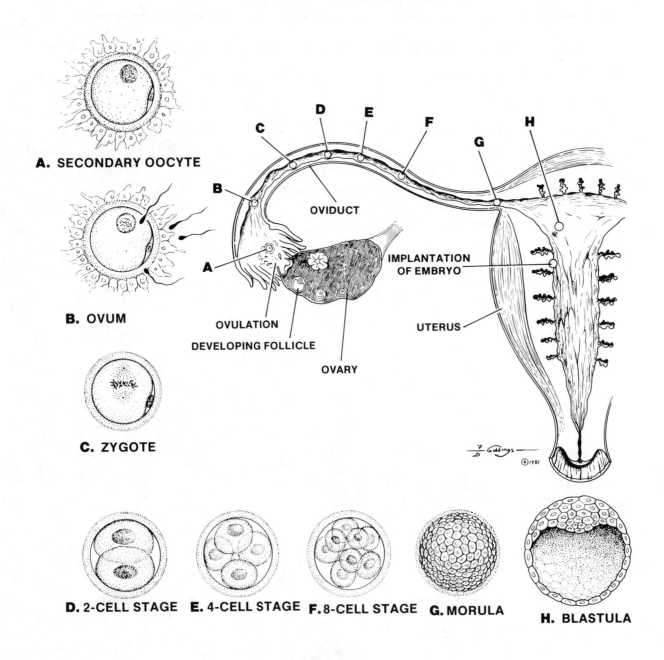

A. SECONDARY OOCYTE

B. OVUM

C. ZYGOTE

OVIDUCT

OVULATION

DEVELOPING FOLLICLE

OVARY

IMPLANTATION OF EMBRYO

UTERUS

D. 2-CELL STAGE   **E.** 4-CELL STAGE   **F.** 8-CELL STAGE   **G.** MORULA   **H.** BLASTULA

FIGURE 29.19
DIAGRAMMATIC REPRESENTATION OF THE MOVEMENT OF THE OOCYTE & EMBRYO

The blastocyst is equivalent to the blastula stage; the blastocyst cavity does not have a wall of uniform thickness: a cluster of cells occurs at one end. The wall of the blastocyst will give rise to the **chorion.** This membrane has two important functions: (1) during early development (first 3 months) it produces **HCG,** the hormone responsible for keeping the corpus luteum functioning; and (2) it initially absorbs nutrients from the endometrium through villi and later gives rise to the fetal portion of the **placenta.** The cluster of cells is known as the **inner cell mass,** which will give rise to 4 structures: **embryo, amnion,** which forms a protective, fluid-filled cavity around the developing embryo; **yolk sac,** present even though the human oocyte has no yolk granules; and **allantois,** whose most important function is probably the formation of the umbilical blood vessels. (Refer back to figure 29.18 for the structures printed in boldface.)

A few days after entering the uterus, the blastocyst undergoes the process of **nidation** (implantation), during which it buries itself in the endometrium from which it will acquire nutrients for the rest of its development.

Gastrulation occurs in somewhat the same manner as in the chick since a primitive streak is formed. The **embryo** with its 3 germ layers (ectoderm, endoderm, and mesoderm) is formed within 2 weeks after ovulation. By the end of 8 weeks of development, all of the body parts are present and it is now called a **fetus,** rather than an embryo. After this time, the organs develop further to eventually become functional and the fetus grows rapidly in size. To accommodate the increased nutritional demands for this growth, the **placenta** serves as the site of nutrient and waste exchange betweeen fetus and mother.

To give you a better appreciation of the growth involved, the following chart indicates the "crown-rump" length (i.e., sitting height) of the human fetus at different stages of development:

2 weeks — 0.23 mm	5 month — 7 in.
1 month — 3/8 in.	6 month — 9 in.
2 month — 1 in.	7 month — 11 in.
3 month — 3 in.	8 month — 14 in.
4 month — 5 in.	

Examine the materials on display in the lab. The models display the relationship of the fetus to the uterus, and the charts illustrate the process of birth.

VIII.   Review Questions

A.   Why is the maintenance of the corpus luteum so important during human pregnancy?

_____

_____

_____

B.   What are the maternal and fetal portions of the placenta? _____

_____

C.   How do substances actually get from the blood stream of the mother to the child? _____

_____

_____

D.   The pregnancy test is based on the presence of what hormone? _____

E.   List the 3 germ layers and name two structures which develop from each.

_____

_____

_____

F.   Define the following terms.

   1.   gastrulation —

   2.   morphogenesis —

   3.   fertilization —

   4.   cleavage —

   5.   blastopore —

   6.   archenteron —

   7.   invagination —

   8.   blastocyst —

   9.   nidation —

   10.   neuralation —

G.   How does embryonic nutrition in the chick differ from that in a mammal? _____

_____

H.   Examine the placenta. Do the chorionic villi entirely surround the embryo or are they

confined to an area of attachment? _____

I.   The chemical composition of the amniotic fluid is similar to that of blood. Is this of any

significance? _____

_____

J.   How does embryonic respiration differ in the chick and human embryos? _____

_____

K.   At what age do you begin to see the formation of eyes in the chick embryo? _____

L.   At about what age do you see development of segmentation of the chick embryo body?

_____

EXCRETORY SYSTEM

exercise **30**

I. Objectives

After the completion of this exercise, the student should be able to do each of the following:

A. Identify the parts of the human urinary system.

B. Identify the structural components of the nephron and describe where they are located.

C. Describe where filtration and reabsorption occur in the nephron and explain what is involved in each process.

D. Explain why increased salt intake lowers the rate of urine formation.

E. Describe the relationship between the rate of urine formation and specific gravity (from lab discussion of results).

F. Explain exactly how alcohol or caffeine increases urine output (applicable if lab instructor uses test solution E).

G. Describe the normal range for the pH of the urine and list several factors that might alter the pH.

H. Answer the review questions at the end of this exercise.

II. Introduction

The amount and chemical composition of the body fluids are controlled by **input** (food, fluid, etc.), **metabolism** and **output** (urine).

The **kidneys** control output because they regulate the cellular environment by selective excretion of those substances in excess or toxic to the body and by a reabsorption of those substances beneficial to the body.

In this lab we will study the anatomy of the kidney and also experimentally test the effect of intake of different substances on kidney function.

III. Kidney Anatomy

Each kidney is composed of approximately 1 million microscopic units known as **nephrons.** These tubular structures are the functional units of the kidney. Each one is associated with two capillary networks. Because of the pressure of blood in the **glomerulus** (first capillary network), dissolved substances and water are filtered out of the blood into the cavity of the **Bowman's capsule.** This solution in the capsule is basically blood plasma minus plasma proteins. When the solution reaches the tubular portion of the nephron, certain of the dissolved substances and most of the water are reabsorbed into a second capillary network and re-enter the bloodstream. Specifically, sodium is reabsorbed into the capillaries from the proximal convoluted tubule by active transport. As more sodium is reabsorbed, the water concentration in the capillaries drops, allowing water from the tubule to move into the capillaries by osmosis. Thus, if more sodium is available to be reabsorbed, then more water will enter the capillaries by osmosis. Thus, a combination of **filtration** and **reabsorption** maintains the proper balance of water and salts within the body and also rids the body of metabolic wastes (especially nitrogenous wastes). Fluid remaining in the tubules after reabsorption is **urine,** the color of which is dependent on the concentration of bile compounds present. The fluid from all of the nephrons drains into a common cavity.

The kidney is divided into three general regions: **cortex, medulla,** and **pelvis.** The cortex is the outer region of the organ, containing mainly the glomeruli and convoluted tubules of the nephrons. The medulla has pyramid-shaped groupings of collecting tubules and Henle's loops. The pelvis is the region into which the contents of the nephrons are emptied. From this point the urine is drained into the **urinary bladder** via a **ureter.** After temporary storage, urine is passed to the outside via the **urethra.**

Observe the charts, models, and preserved sheep's kidney on display, using Figure 30.1 as a guide.

IV. Urinary Physiology

Analysis of urine can yield valuable information about the condition of the body. Certain diseases are characterized by the presence of substances in the urine which normally should not be present. Presence of glucose, for example, indicates that there is so much glucose in the bloodstream that the kidney tubule is incapable of completely reabsorbing it, and is indicative of diabetes mellitus. Presence of ketones in urine is also a symptom of diabetes. These chemicals result from excessive metabolism of lipids due to the cells' inability to obtain and use glucose. Presence of blood or protein may indicate damage to the urinary tract by a bacterial infection. Volume of urine produced indicates the level of fluid intake. A small amount of highly concentrated urine indicates a state of body dehydration. The kidney has the ability to adjust the pH of the urine over a wide range. It can excrete or reabsorb ions to maintain a constant blood pH of 7.4. A drastic change in pH may indicate malfunction in the nephron. Thus, analysis of urine composition as well as of volume is a valuable tool in determining the general state of health of the individual as well as the specific condition of the kidneys.

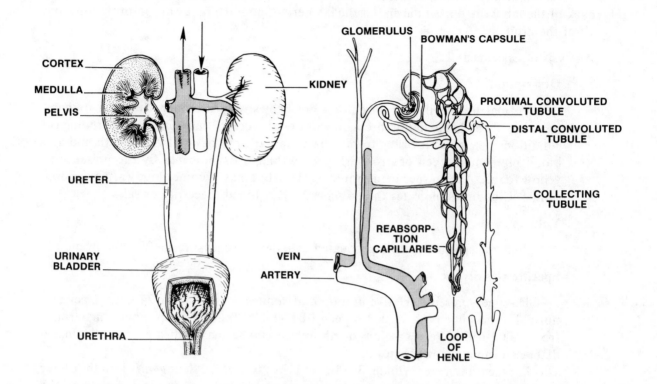

FIGURE 30.1 — THE URINARY SYSTEM AND THE NEPHRON

In the following experiment we will demonstrate the effect of intake of various substances on the volume and composition of urine produced.

A. Experimental Procedure

At least 4 test groups will be set up (more, if the instructor so desires). Volunteers should be chosen in the previous lab period so that their food and water intake can be restricted for two hours before the experiment. Record the last time of urination prior to the experiment. At the beginning of the lab period, each subject should empty the bladder as completely as possible. This urine should be saved and labeled with the subject's name. This specimen will serve as the **control.**

Each subject will drink **one** of the following solutions. Drink only as much of the test solution as is comfortably possible.

Solution A: 700 ml distilled water
Solution B: 700 ml 0.9% NaCl
Solution C: 250 ml 1.5% NaCl
Solution D: 400 ml 1.0% $NaHCO_3$
Solution E: 500 ml beer, coffee, or cola beverage

Begin timing the subject as soon as the fluid is consumed. A urine specimen should be collected *every 20 minutes* for a total period of *two* hours. On the data sheet for the test solution to be analyzed by your group, indicate the times at which the specimens are to be collected. All specimens, including the control, will be analyzed separately.

Note: No other fluids should be consumed during the experimental period. IF A SUBJECT IS UNABLE TO PRODUCE ENOUGH URINE AT ANY OF THE COL-LECTING PERIODS, THE URINE SHOULD BE RETAINED AND COMBINED WITH THE NEXT COLLECTION.

Record all the information for your test solution on the appropriate data sheet at the back of the lab write-up. At the end of the lab period, pool the necessary information from all of the groups.

B. Analysis of Specimens

1. Description

Normal urine will vary in color from light straw to amber, due to a pigment resulting from hemoglobin breakdown. Certain colors may indicate pathological conditions — e.g., milky (due to pus, bacteria, fat), red or smoky brown (blood and blood pigments), green or brownish yellow (bile; also indicated by the presence of yellow foam). A fresh specimen should be clear, but may become cloudy after standing awhile. Cloudiness indicates the presence of pus, blood, bacteria, or salts.

2. Volume

Pour the specimen into a graduated cylinder and measure the volume in ml.

3. Specific Gravity

This measures the relative amounts of solutes present. In a 24-hour specimen, normal urine will have a value between 1.015-1.025. Individual specimens may range from 1.002-1.030. Values below or above this range may indicate kidney damage, diabetes insipidus, or diabetes mellitis.

Fill the urinometer cylinder 3/4 full of urine. Lower the float *gently* into the center portion of the cylinder, being sure not to touch the sides of the cylinder. Read the specific gravity value from the level indicated by the bottom of the urine meniscus. Be sure to wash, rinse, and dry the apparatus carefully after *each* use.

4. pH

Freshly voided urine has a normal pH range of 4.8-7.5. If the **pH** is very low, this may indicate acidosis, fever, or high protein diet; if the **pH** is very high, this may indicate urine stagnation in the bladder, anemia, or cystitis.

To use pH indicator paper, dip into urine 3 times. Tap off excess urine and, after the proper interval of time, compare color with the color chart. To use the Corning pH meter, follow these directions:

(a) With the knob of the machine on "standby" (STDBY) lift the electrodes from the water and wipe them dry with Kim-wipes.

(b) Lower the electrodes into a buffer of known pH.

(c) Turn the top knob to the "pH" position and turn the calibration knob (lower knob) until the needle on the scale is set for the pH of the buffer (usually 7, but ask your instructor).

(d) Place the top switch back to "standby"; lift the electrodes and place them in water to wash off the buffer.

(e) Dry with Kim-wipes and then place electrodes in the urine sample.

(f) Turn the knob to "pH" and read sample pH from the scale.

(g) Turn knob back to STDBY, then lift the electrodes and immerse again in water.

CAUTION: Always leave top knob in STDBY position unless you are actually testing the buffer or the urine. NEVER remove electrodes from buffer or urine when knob is in the "pH" position!

Do *not* adjust the middle knob (temperature control). Electrodes should be immersed in *water* when you are not using the machine.

5. **Chloride Estimate**

Sodium chloride (NaCl) is the main form of chloride present in the urine and accounts for about half of the inorganic substances excreted through the urine.

Measure 10 drops (1/2 ml) of urine into a test tube. Add 1 drop of 20% potassium chromate (K_2CrO_4). Using the labeled dropper, add 1 drop at a time of 2.9% silver nitrate ($AgNO_3$) solution, shaking the test tube constantly. Count the number of drops necessary to change the solution from yellow to brownish tan. Each drop of silver nitrate represents about 1 g of NaCl per liter of urine.

a. Record number of drops used under NaCl g/1.

b. Determine the total grams NaCl in the entire sample using the following equation:

$$\text{Total NaCl (grams)} = \frac{\text{ml of sample}}{1000 \text{ ml}} \times \text{Number of drops AgNO}_3 \text{ used}$$

The first figure for chloride estimate (g/1) is based on a one liter volume even though the sample voided is much less than this. The actual amount of salt excreted (total NaCl) depends on what fraction of one liter is voided at a particular time (ml of sample / 1000). Measurement in g/1 gives a relative figure that can be compared with other samples and among subjects, whereas "total NaCl" is a measurement of how much salt was physically present in a given sample.

6. **Presence of Glucose** (control sample only)

To make a "pathological" urine sample, add 10 drops of a glucose solution to 5 ml of the urine sample in a test tube. Use 5 ml of the urine sample without the added glucose in another tube to represent the normal condition. Add 5 ml of Benedicts Reagent to each tube and heat both of them in a water bath. The appearance of an orange to brick-red color is an indication that glucose is present. Was there any glucose in the normal control sample?

7. **Heller Ring Test** for the Presence of Albumin (control sample only)

Make a "pathological" sample by adding 10 drops of albumin solution to 5 ml of the urine sample in a test tube. As before, use another tube with 5 ml of urine without albumin added to serve as a control. Place 2-3 ml of concentrated nitric acid into 2 empty test tubes. Slant the tube and carefully pour 2-3 ml of the normal and pathological urine down the sides of each tube. Two distinct layers should form and the contents of the tubes should not be mixed. The formation of a white ring at the junction of the two liquids indicates the presence of albumin.

V. Review Questions

A. During the 2 hour period was more fluid excreted than was consumed by any of the subjects? Which ones? Why would this be so?

B. What were the differences in the results from the intake of the hypotonic, isotonic, and hypertonic solutions? (Consider only solutions A, B, and C.)

C. Did bicarbonate solution cause more, or less urine output than did water? What effect did this solution have on the pH of the urine?

D. When would the urine of a normal individual contain glucose?

E. Why is the presence of blood or protein in urine considered abnormal?

F. Compare the location of the proximal convoluted tubule to that of the distal convoluted tubule.

340

DATA SHEET A — SOLUTION: _____

Volunteer's Name _____

Time of last voiding before lab period _____

Volume of test substance drunk _____

Urine Specimen	Time	Volume Voided (ml)	Rate of Urine Formation (ml/min)	Specific Gravity	pH	NaCl g/l	Total g NaCl	Description of Sample
Control								
1 (20 min)								
2 (40 min)								
3 (1 hour)								
4 (1 hour, 20 min)								
5 (1 hour, 40 min)								
6 (2 hours)								

Conclusions:

DATA SHEET B — SOLUTION: _____

Volunteer's Name _____

Time of last voiding before lab period _____

Volume of test substance drunk _____

Urine Specimen	Time	Volume Voided (ml)	Rate of Urine Formation (ml/min)	Specific Gravity	pH	NaCl g/l	Total g NaCl	Description of Sample
Control								
1 (20 min)								
2 (40 min)								
3 (1 hour)								
4 (1 hour, 20 min)								
5 (1 hour, 40 min)								
6 (2 hours)								

Conclusions:

DATA SHEET C — SOLUTION: _____

Volunteer's Name _____

Time of last voiding before lab period _____

Volume of test substance drunk _____

Urine Specimen	Time	Volume Voided (ml)	Rate of Urine Formation (ml/min)	Specific Gravity	pH	NaCl g/l	Total g NaCl	Description of Sample
Control								
1 (20 min)								
2 (40 min)								
3 (1 hour)								
4 (1 hour, 20 min)								
5 (1 hour, 40 min)								
6 (2 hours)								

Conclusions:

DATA SHEET D — SOLUTION: _____

Volunteer's Name _____

Time of last voiding before lab period _____

Volume of test substance drunk _____

Urine Specimen	Time	Volume Voided (ml)	Rate of Urine Formation (ml/min)	Specific Gravity	pH	NaCl g/l	Total g NaCl	Description of Sample
Control								
1 (20 min)								
2 (40 min)								
3 (1 hour)								
4 (1 hour, 20 min)								
5 (1 hour, 40 min)								
6 (2 hours)								

Conclusions:

DATA SHEET E — SOLUTION: _____

Volunteer's Name _____

Time of last voiding before lab period _____

Volume of test substance drunk _____

Urine Specimen	Time	Volume Voided (ml)	Rate of Urine Formation (ml/min)	Specific Gravity	pH	NaCl g/l	Total g NaCl	Description of Sample
Control								
1 (20 min)								
2 (40 min)								
3 (1 hour)								
4 (1 hour, 20 min)								
5 (1 hour, 40 min)								
6 (2 hours)								

Conclusions:

ECOLOGY

I.   Objectives

After the completion of this exercise, the student should be able to do each of the following:

A.   Define the term "ecosystem".

B.   Describe the creation of a dune ecosystem near a sandy seashore and the changes which occur during ecological succession in such an area.

C.   List the 6 ecosystems (zones) in and adjacent to a sandy seashore area and list characteristic plant and animal species which inhabit each zone.

D.   Explain the operation of primary plant succession in the swamp and ridge subzones of the dune-based forest.

E.   Describe the different trophic (feeding) levels in a forest ecosystem, and explain how they depend upon each other.

F.   List plants which belong in the canopy, understory, and shrub-ground layer of the forest.

G.   Describe several ecological interactions among organisms which are important to the functioning of the forest ecosystem.

H.   Answer the questions throughout this exercise.

II.   Introduction

The purpose of this lab is to observe and analyze several ecosystems found in and adjacent to a sandy seashore. An **ecosystem** is an ecological unit which is composed of a number of populations of organisms which interact with each other and with their physical environment.

In any ecosystem, the primary source of energy is sunlight (there are a few exceptions to this statement), the energy of which is trapped in food molecules produced by photosynthesis by organisms having chlorophyll. Such organisms, mostly plants and some protists and monerans,

347

are referred to as **producers.** All other organisms must get their energy by feeding on producers or their remains, or on other organisms which feed on producers. Organisms which feed directly on the producers are known as **primary consumers.** This is one of several **trophic** (feeding) **levels** which exist within an ecosystem. Organisms which feed on primary consumers are known as **secondary consumers,** and so forth. In the end, all organisms or their dead remains are decomposed for energy by such **decomposer organisms** as bacteria and fungi.

The chemicals making up the bodies of organisms can then be recycled, but the energy is finally lost to the ecosystem in the form of unusable heat energy. Thus energy does not cycle in an ecosystem, but makes a one-way trip through it.

Usually, the total amount of **biomass** is reduced with each successive trophic level because of inefficiency of energy transfer and heat loss at each exchange. Therefore, one would expect the biomass of producers in an ecosystem to be greater than the biomass of primary consumers, and the biomass of primary consumers to be greater than the biomass of secondary consumers.

In one portion of this lab, we will describe the producers, consumers, and physical environments of several ecosystems, and then later in the lab, we will investigate the functional relationships among these parts. First, let us take a look at a sandy seashore and its adjacent areas in general and examine some of the ecosystems (zones) found there.

For millions of years, bay areas have been formed, washed away, and reformed through the combined action of sand, ocean waves, and winds. With each cycle of disappearance and reappearance, the type of land and sea life varied. The currently existing sandy coastal areas have been forming over the past 20-25,000 years. Sand has been continuously deposited along the shoreline, extending the land farther and farther into the bay areas. The sand left behind by the receding ocean has been formed into sand dunes, which gradually, over hundreds of years, became invaded by plant life from farther inland as the intensity of salt spray decreased. Plant communities of increasing complexity developed: beach dunes converted to grass lands, then to dune scrub areas, then to complex forests. Low areas became pools, swamps, or marshes. With the change in plant life came the accompanying changes in animal life, adapting to the environmental changes.

III. Ecosystem Zones

Six ecosystem zones are usually found more or less parallel to the shoreline of a sandy shore. Certain parts of this area will not fit into one of the six zones because the zones plus the intermediate stages between zones demonstrate the concept of "ecological succession" in which there is no clearcut boundary between the zones, but rather a continuous transition from one type of community to another. The shore is dynamic, always changing in its plant and animal life. Use the diagram in Figure 31.1 and the discussion above as guides to understanding these zones better.

A. Beach Zone

The **intertidal zone,** or **lower beach** area, is exposed only at low tide. Because of the tremendous variation in conditions — waves, scorching sunlight, and nighttime coolness —no vegetation can live here. The only animals which can survive here are those which burrow into the sand and are filter feeders or intermittent scavengers.

What is the primary source of food energy in this zone?

List the organisms found here and try to determine their trophic levels.

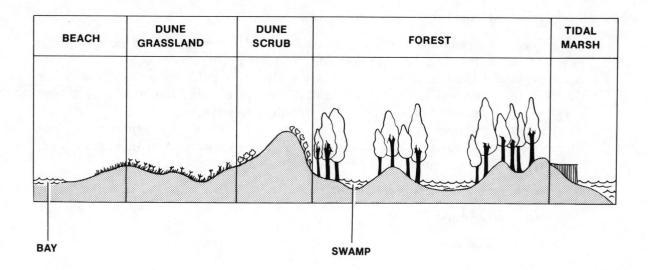

| BEACH | DUNE GRASSLAND | DUNE SCRUB | FOREST | TIDAL MARSH |

BAY

SWAMP

FIGURE 31.1
DIAGRAMMATIC CROSS-SECTION THROUGH A SEASHORE

The **middle beach** is that area of the beach stretching between the high tide mark and the first sand dunes with vegetation. Here will be found the debris left behind by the ocean tides — sea shells, egg cases, ocean plants, driftwood, and washed-up bodies of ocean animal forms. Some of the animals here live in homes tunnelled into the sand and scavenge dead animal material not already claimed by the shore birds. They are often protected in their environment by the color pattern simulating the sandy coloration of the shore.

Birds inhabit the beach area looking for food wherever it can be found. Some dive into the waves head-first to grab fish; others fly along the surface of the water, skimming small fish into their lower bills; and on the beach, some dodge incoming waves in their search for dead animal matter or insects; while others wander about on the middle beach looking for food.

What is the primary source of food energy in this zone?

List the organisms found here and try to determine their trophic levels.

The **upper beach** is the area where vegetation first really establishes itself. Exposure to salt spray and wind permits only grasses to exist here. Not only do they serve to anchor the sand dunes, but they also protect areas behind the beach from wind erosion, wave action, and salt spray. Living among the grasses are various invertebrates.

What are the producers in this zone?

List some primary consumers seen here.

Are there any secondary consumers that you can see?

What would be the eventual expected result of building houses and hotels in this zone?

349

B. Dune Grassland Zone

This zone is represented by the ridge of sand dunes. It is separated from the upper beach by a sandy bluff, marking the inner boundary of erosion caused by recent storms; beyond this area, inland plants and animals are not subject to regular invasions of sea water. However, the winds and salt spray still control the type of vegetation able to exist here, bringing in new sand and handicapping the growth of woody plants.

The most common plants in this zone are still grasses, although they are now interspersed with some hardy herbaceous plants. Again, various types of animals are found in this region — mostly insects and animals which feed upon insects.

What are the producers in this zone?

Primary consumers?

Secondary consumers?

C. Dune Scrub Zone

This zone is located directly behind the dune grasslands and is lower in elevation than the ridge forming the dunes. It is characterized by the presence of woody plants which are able to grow in this area due to the decreased amount of salt spray reaching them and to the increased stability of the soil. However, periodic increases in the amount of salt spray during storms controls the height of the plants, creating a bushier type of growth. In this area, woody thickets are interspersed with more open spaces containing a variety of grasses, vines, and sometimes, cacti.

Between the sandy hills with their woody plants may be some low lying grassy meadows in which the soil is deep and relatively moist. It is in these areas that different species of rushes, grasses, weeds, and wildflowers may be found. Why would the soil be deeper here?

Because of the abundance of plant life, a variety of animals will also be found in the dune scrub zone. Birds of various kinds inhabit the trees. In addition, many types of reptiles and amphibians (toads, lizards, skinks, and snakes) as well as a whole range of insects and arachnids inhabit this zone. Few mammals actually live in this region, though occasionally mice, squirrels, rabbits, etc., may be seen.

List several species of producers found in this zone.

Which type of producers predominate (think in terms of biomass)?

List several species of primary consumers found here.

Secondary consumers?

D. Forest Zone

This is the most complex zone found adjacent to a sandy beach. It is located between the dune scrub zone near the beach and the tidal marsh zone along the bay. The forest zone itself encompasses a variety of areas, each of which differs and has its unique characteristics. Each subzone differs in the type of vegetation growing among the dominant trees, usually pines. There are a larger number of animal species here than in any other ecosystem associated with the sandy beach.

In the forest zone of southeastern sandy beach areas in the United States, an area of rolling sandy hills is usually found populated with pines and hardwoods, predominantly oak. These hardwoods form the **first story** of the forest **canopy.** Above these trees rise the crowns of the pines which form the **second story** of the canopy.

Parallel to the beach is a ridge of high sand dunes which make up the **great dune subzone.**

The most highly developed plant community in the area is the **ridge subzone**. Again, in temperate southern beach areas this is usually a mixed pine-hardwood kind of community, although in any one spot, one species of tree may be prevalent over others.

Note that the larger trees grow at the bases of the ridges. Below the canopy is an **understory** formed by other shrubbier trees. And below this will be found various small shrubs and herbs bearing flowers and fruits, in addition to a variety of ferns, mosses, fungi, lichens, and algae.

Frequently, a **swamp subzone** will be found within the forest. This is populated by cypress trees in southern beach areas. Located between dune ridges, these swamps vary only slightly in their vegetation due to differences in water level. In the deeper swamps, cypress may be the only trees present, while in shallower areas, trees whose roots can tolerate mucky soil may accompany them. Hanging on the branches of these swamp trees may be found certain epiphytes, or plants which grow on other plants to gain access to sunlight. In the southern beach areas, one of the commoner epiphytes is spanish moss, a flowering plant related to the pineapple. Protruding from the swamp waters are the knobby cypress knees, part of the trees' root systems. Around these knees, material may accumulate over the years, resulting in the formation of islands. Many woody and herbaceous plants grow on these tiny islands as well as along the shallow edges of the swamp.

In addition to the forested or grassy areas of the forest zone is an **open water subzone.** There are frequently a number of these areas, consisting of pools or small lakes which differ from the cypress swamps. Some are surrounded by cattails or rushes, and water lilies can be found on some of them.

Name the predominant producer species in each of the subzones of the forest zone which you observed.

List the most prevalent primary consumers in each of the subzones which you observed.

List the decomposers identified.

Is the animal life in the forest subzone layered in the same way as is the plant life? Explain, using examples.

351

E. Tidal Marsh Zone

In the protected tidal areas of a sandy beach area, sandy muck plains of grasses extend like shelves into the tidal streams and embayments. These constitute the **tidal marsh zone.** The predominant types of plants found here are reeds and salt marsh grasses. Closest to the forest, these are mixed with water tolerant and salt tolerant herbaceous dicots. Living among the grasses are fiddler crabs, snails, muskrats, and various insects and crustaceans. Buried in the mud are mussels, oysters, mud snails, and sea squirts. Flying about the grasses are different types of insects and birds, especially waders, which feed on the other animals mentioned. The marsh acts as the first step in the food webs leading to most life in the bay itself. Breakdown of grasses puts food into the water for a large variety of protozoan and planktonic life which act as food for larger water dwellers.

What are the main producer organisms in the tidal marsh zone?

Name the species of animals which are primary consumers here?

Secondary consumers?

What evidence is there of the activities of decomposer organisms?

F. Roadside Zone

This zone represents all those areas where man has removed the natural vegetation and replaced it with another plant community. This vegetation is complex but unstable, in that it changes yearly.

Keep in mind that all the zones mentioned, the forest zone in particular, contain a tremendous variety and number of animals, even though the zones are described primarily according to the dominant plant types. Most of these animals can be seen at some time during the day, depending on the observational ability of the student. Take a good look at the organisms you find as you walk along; they provide a good opportunity to study the ecosystem from several points of view.

IV. Review Questions

A. Which of the ecosystems that you have seen is the most diverse (has the largest number of different species)? _____

B. Which ecosystem is the least diverse (has the fewest number of different species)? _____

_____

C. If you eliminated the single most abundant species in each of the two ecosystems above (by logging, fire, air pollution, heavy picnic use, etc.) which of the ecosystems would be affected the most? Which ecosystem would be changed the least? _____

_____

D. Does your answer to C support or disagree with the hypothesis that stability is correlated with diversity? _____

E. If you cut or burned or polluted out much of a terrestrial ecosystem, what would you expect the effect on adjacent aquatic systems to be? _____

_____

_____

F. The canopy of a forest is defined as consisting of those trees whose crowns are at the top forming a continuous layer of the forest. What were the major canopy trees in the ecosystem you studied? _____

G. The understory is composed of those plants whose diameters measure two inches at the height of 4.5 feet above ground level and whose crowns are below the top layer of the forest. List the major understory trees that you saw in this exercise. _____

_____

H. The shrub and ground layers are composed of those tree seedlings and saplings less than 5 cm. in diameter. Shrubs and herbaceous plants would also be in this layer. List several plants that you saw in this layer. _____

_____

I. Were the same canopy tree species represented as seedlings or small trees in the lower layers?

_____

J. What do you think would be the future course of plant succession in this forest? _____ _____ Why? _____

_____

K. Along the way, your instructor pointed out a fallen tree, which was being decomposed by bacteria and fungi. What effect should this decay process have on (1) the mineral content of the soil? _____

(2) the amount of stored energy within the ecosystem? _____

(3) possible changes within the ecosystem? _____

_____

POPULATION GENETICS AND EVOLUTION

I. Objectives

After completing this exercise, the student should be able to do each of the following:

A. Explain the Hardy-Weinberg Law in terms of gene frequencies in a population.

B. State five conditions that must be met in order for the Hardy-Weinberg Law to operate.

C. Calculate the proportion of AA, Aa, and aa genotypes in a population if the frequencies of "A" and "a" are known.

D. Describe the effect of natural selection on stable gene frequencies.

E. Explain the relationship between **natural selection, differential reproductive rates,** and **the adaptation of a population.**

F. Discuss adaptations of plants to dry desert-like environments, aquatic environments, and competition for light.

G. Explain the advantages of the carnivorous habit to plants such as the pitcher plant and the Venus flytrap.

H. Define and give examples of: **mimicry, convergence** (convergent evolution), **divergence,** and **dispersal mechanisms.**

I. Answer the review questions at the end of this exercise.

II. The Hardy-Weinberg Law

Often a trait will be very rare even though it is caused by the presence of a dominant allele. For example, the trait of having 6 fingers on each hand is controlled by a dominant allele but this trait is rare in the human population. You may wonder why this trait does not become more common, and also perhaps why such recessive traits as albinism and colorblindness do not disappear entirely.

The Hardy-Weinberg Law states that under certain conditions of stability, both the gene frequencies and the genotype frequencies **remain constant** from generation to generation in a sexually reproducing population. This law applies regardless of whether the gene's expression is dominant or recessive. For example, the allele for having 6 fingers has a certain low frequency which does not vary noticeably and thus the trait remains rare.

This genetically stable condition is maintained under the following conditions: (1) a large population; (2) no mutations; (3) no organisms entering or leaving the population; (4) random mating; and (5) no natural selection.

To test the principles of the Hardy-Weinberg Law, the following experiment will be set up. An artificial population will be used to symbolize a group of sexually reproducing animals in which there are equal numbers of males and females. Some individuals of this population are homozygous for the dominant allele of a gene, some are homozygous recessive, and some are heterozygous. Let 'p' symbolize the frequency of the dominant allele in the population (p can vary from zero to one), and let 'q' symbolize the frequency of the recessive allele. Note that $p + q = 1$, so if p is already known, the $q = 1 - p$. It can be proved that if the population is very large and if there is no selection either for or against any particular genotype, the genotypes should be maintained in the population in the following proportions:

Proportion of the population that is homozygous dominant (AA) $= p^2$ (pp)

Proportion of the population that is heterozygous (Aa) $= 2pq$

Proportion of the population that is homozygous recessive (aa) $= q^2$ (qq)

To test if this relationship holds for a population of 64 breeding individuals in which there are equal numbers of recessive and dominant alleles ($p = q = \frac{1}{2}$), and where the parents are obtained by chance or at random, perform the following experiment.

Place 16 red toothpicks in a container. These represent the homozygous dominant individuals (AA). Sixteen red toothpicks are placed in the container because if $p = \frac{1}{2}$, then $p^2 \times 64 = 16$. Place 32 black toothpicks (Aa individuals) in the container. Why use 32 toothpicks?

Finally, place 16 yellow toothpicks (aa individuals) in the same container. Shake up the container and draw at random two toothpicks. These two toothpicks will be considered as parents which will produce 4 offspring, and the genotypes of the parents will determine the genotypes of the offspring according to genetic laws. The possible types of matings along with the kind of offspring which can be produced by such parents are as follows:

red (AA)	× red (AA)	=	4 red (AA)
yellow (aa)	× yellow (aa)	=	4 yellow (aa)
black (Aa)	× black (Aa)	=	1 red (AA), 2 black (Aa), 1 yellow (aa)
red (AA)	× yellow (aa)	=	4 black (Aa)
red (AA)	× black (Aa)	=	2 red (AA), 2 black (Aa)
yellow (aa)	× black (Aa)	=	2 yellow (aa), 2 black (Aa)

Therefore, if you should happen to draw two black toothpicks, place one red, 2 black, and one yellow toothpick in a pile which represents the individuals making up the next generation. Replace, in the container the toothpicks drawn as **parents** so that the probabilities of drawing will not be affected by removal of the toothpicks. Shake the toothpicks and draw a second pair which will also have 4 offspring according to the parents' colors. Continue the same procedure until the F_1 generation is composed of 64 offspring (by drawing 16 pairs of toothpicks). Record the composition of the F_1 generation in Table I. Repeat the process, drawing from the F_1 generation to obtain the parents of the F_2 generation, these parents reproducing in the same way as their parents. Continue through the formation of the F_4 generation. Calculate the frequency of the recessive allele in each generation by doubling the number of yellow toothpicks and adding this to the number of black toothpicks (why is each black toothpick counted only once?). Record this in Table I.

TABLE I

Population Size is 64; $p = q = \frac{1}{2}$
Each Mating Produces 4 Offspring

GENERATION	NUMBER OF REDS (AA)	NUMBER OF BLACKS (Aa)	NUMBER OF YELLOWS (aa)	FREQUENCY OF RECESSIVE ALLELE
original	16	32	16	64/128
F_1				
F_2				
F_3				
F_4				
Class Average of F_4				

Has the original ratio of genotypes been substantially changed even though the breeding population is small?

The above experiments assumed random mating. Do you think that humans mate at random? List some of the factors which might disturb the randomness of human mating.

III. Natural Selection

Natural selection occurs when a certain trait is better or worse suited to a particular set of environmental conditions. This selection would tend to favor a particular genotype which produces a trait that is better suited to the environment, and would work for the elimination of a genotype which produces a trait that is not well adapted to its particular environment by allowing higher reproductive rates for the former than for the latter. As mentioned in part II, natural selection will tend to upset the stable gene and genotype frequencies and work against the Hardy-Weinberg equilibrium. Since evolution occurs when natural selection works upon random variations in a population, it upsets the stability predicted by the Hardy-Weinberg Law.

To demonstrate natural selection, perform an experiment where no offspring are produced by homozygous recessive individuals. This is equivalent to selection against the recessive trait.

Make up the parent population as before. Draw at random two toothpicks from the container. Discard any yellow toothpick that is drawn, and draw again to replace it. Add 4 offspring to the next generation according to the genotypes of the parents as in the last experiment. There will be no yellow parents in this experiment but one of the offspring will be yellow when both parents are black (Aa). Continue drawing pairs until there are 64 individuals in the F_1 generation. Record the results in Table II. Repeat the process through the formation of the F_4 generation.

357

TABLE II

Population Size is 64; $p = q = \frac{1}{2}$;
Each Mating Produces 4 Offspring
Homozygous Recessive Individuals (aa) Do Not Mate

GENERATION	NUMBER OF REDS (AA)	NUMBER OF BLACKS (Aa)	NUMBER OF YELLOWS(aa)	FREQUENCY OF THE RECESSIVE ALLELE	CLASS AVERAGE OF FREQUENCY
original	16	32	16	64/128	
F_1					
F_2					
F_3					
F_4					

Why is it so difficult to eliminate an inherited condition from a population of animals if the phenotype is determined by the homozygous condition of a recessive allele?

If all the individuals showing the phenotype determined by a dominant allele were eliminated (or prevented from reproducing) how many generations would it take to eliminate this allele from the population?

IV. Optional: Computer Simulation

The preceding "toothpick" demonstration of the Hardy-Weinberg Law picked 16 pairs of parents from a population of 64 organisms which would be considered a small population. Biologically, populations may have a far greater number of members. Demonstrating the law with this number of toothpicks would be very impractical! Fortunately, a computer program can be written to do the same thing that was done with the toothpicks but the computer can choose 160 or even 1600 sets of parents in only a fraction of a second longer than it can pick 16 sets. The secret of the program is a "random number generator" which picks pairs of numbers at random (like pairs of parents or toothpicks). The computer uses the position of the numbers rather than colors of toothpicks to determine the genotypes of the parents. For example, in the original population in Table I, the computer would call numbers from 1 through 16 "AA", numbers 17 through 48 "Aa" and numbers 49 through 64 "aa". Once the computer has chosen 2 numbers (converted to genotypes of parents) the second part of the program simply determines the genotypes of their offspring, and repeats the process over and over until a certain number of parents are selected just like the toothpicks demonstration.

There are 2 programs available which perform the same function as the 2 toothpick experiments. HWLAW gives a simulation of a population according to the Hardy-Weinberg Law, and NATSEL demonstrates how natural selection works on gene frequencies. The student can call up either program and can plug in different values for gene frequencies (p and q), size of the population, number of parents to be selected at random (always ¼ of the population size), and number of generations to be calculated.

Your lab instructor will give you directions for using the computer terminals on this campus.

The technique of showing how biological events might occur using a computer is called computer simulation or modeling. In the past few years, modeling of entire ecosystems has become a very important area of research since this will allow ecologists to predict in advance what would happen if an ecosystem is modified.

V.   Adaptation

Modifications which enable an organism to live in an environment are known as **adaptations**. Since these modifications arise as a result of natural selection working on favorable variations in a particular environment, these adaptations are the direct result of evolution.

In the following, some of the more noticeable adaptations and evolutionary concepts are discussed, and examples of each are given.

A.   Adaptations to Desert-Like Habitats

Desert plants of various families possess structural devices for collecting and retaining water. Examine the cacti and other succulents (Euphorbes, *Stapelia,* Aloe, etc.) on display.

In the cacti, the spines are modified leaves. What adaptive advantages do these leaves offer the cactus plant?

Also in cacti, the green, fleshy pods are modified stems. What functions do they serve?

Another dweller in desert-like environments is *Tillandsia usneoides* (Spanish moss). This epiphyte has no root system by which water can be obtained. Instead, it absorbs and retains rainwater by means of modified hairs or scales covering its stems and leaves.

Observe the plant, and examine these water-retaining scales under a microscope.

B.   Adaptations to Get the Sunlight

Sunlight is one of the necessary requirements for photosynthesis in plants. So any interference with obtaining sunlight places environmental, and thus evolutionary, pressures on the organism.

Below are listed several approaches which plants have evolved for securing sunlight. Give an example of a plant using each of these approaches:

1.   grows tall _____

2.   grows on other plants _____

3.   climbs _____

4.   grows on the forest floor in the early spring (before trees have leaves on them)

_____

In the adaptation for climbing, plants have used every part — roots, stems and leaves. Some woody plants have formed slender trunks and intertwining leaves and stems to ascend toward the sun. Woody vines are known as **lianas**. Give several examples of lianas.

_____   _____   _____

_____   _____   _____

Other modifications for climbing include **aerial roots** and **tendrils.** Examine the specimens on display and note their adaptations for obtaining sunlight.

C.   **Adaptations to Aquatic Habitats**

Plants must deal with some of the unique differences between terrestrial and aquatic environments in living in or on the water.

List some of the differences between terrestrial and aquatic environments.

_____

_____

_____

_____

Observe some of the aquatic plants on display and list any characteristics which might aid their success in water.

_____

_____

_____

_____

D.   **Adaptations for Living in Nitrogen-Deficient Soil**

Carnivorous plants grow primarily in habitats deficient in nitrogen. These plants secure nitrogen from the bodies of decaying insects and other small animals which have been captured by the plant's modified leaves. This organic nitrogen supply improves growth as compared with plants not furnished a supply of insects. In addition, all carnivorous plants have chlorophyll and hence carry on photosynthesis. Observe the different trapping devices used by the carnivorous plants on display.

E.   **Mimicry and Camouflage**

A major evolutionary advantage to organisms is the ability to escape notice from predators or prey. This is particularly true for animals. One means of doing this is to look similar to the surrounding environment or to some other organism. The former is known as **camouflage,** and the latter, **mimicry.**

Plants sometimes employ mimicry to increase their chances for successful pollination. Examine the specimens on display which demonstrate mimicry and camouflage.

F.   **Divergent Evolution**

Any population of organisms tends to diversify when it spreads (by its particular dispersal mechanisms) to various habitats. These new habitats offer new and unique environmental pressures to the expanding species. This is the force which creates the diversity which may be seen in practically every major group of organisms. The variations within a taxonomic group are examples of **divergence,** or **divergent evolution.**

Examine the members of the Bromeliaceae (pineapple family) and the Euphorbiaceae (spurge family) on display. These families show extreme diversity, even though each maintains its family resemblances (flower structure).

G.   **Convergent Evolution**

Similar habitats may be found in various scattered regions of the world which offer similar environmental pressures to an organism. If two unrelated groups of organisms exist in similar habitats, it should not be surprising to find that the organisms have made similar adaptations in structure and behavior. This tendency of one group of organisms to develop resemblances to another of a different ancestry is known as **convergence,** or **convergent evolution.**

Cacti are found native only in the Western Hemisphere. However, other plant families have also made adaptations to fit arid environments elsewhere. Examine the cacti and euphorbes on display, noting their similarities.

H.   **Dispersal**

The spread of plants from their parents to an area where they can germinate, grow, and mature, is called **dispersal.** Plants have evolved certain adaptations to insure adequate dispersal which can be seen in the fruits and seeds on display. In each case, write down the specific modification of the seed or fruit and tell how it insures seed dispersal.

VI.   Review Questions

A.   List the five conditions under which the Hardy-Weinberg equilibrium is maintained.

_____

_____

_____

B.   If 9% of the population cannot taste PTC (recessive trait), what is the frequency of the non-taster allele in the population? _____ What is the frequency of the taster allele? _____ What percentage of the population will be homozygous dominant? _____ What percentage of the population will be heterozygous? _____ .

C.   Name at least one adaptation found in plants to each of the following:

1.   desert conditions _____

2.   excessive shade at ground level or high population density _____

_____

3.   deficiency of nitrogen in the soil _____

D.   Give an example of convergent evolutin among vertebrates. _____

_____

E.   Give an example of divergent evolution among mammals. _____

_____